Drywall

Level Two

Trainee Guide
First Edition

PEARSON

New York, New York
Columbus, Ohio

NCCER

President: Don Whyte
Director of Curriculum Revision and Development: Daniele Dixon
Drywall Project Manager: Carla Sly and Tania Domenech
Production Manager: Tim Davis
Quality Control Coordinator: Debie Ness
Editors: Rob Richardson and Matt Tischler
Desktop Publishing Coordinator: James McKay
Production Assistant: Laura Wright

NCCER would like to acknowledge the contract service provider for this curriculum:
Topaz Publications, Liverpool, New York.

This information is general in nature and intended for training purposes only. Actual performance of activities described in this manual requires compliance with all applicable operating, service, maintenance, and safety procedures under the direction of qualified personnel. References in this manual to patented or proprietary devices do not constitute a recommendation of their use.

10 9 8 7

PEARSON

ISBN 0-13-604480-8
ISBN 13 978-0-13-604480-2

PREFACE

TO THE TRAINEE

Walk into almost any home, apartment complex, or commercial building and look around. The odds are good that drywall applicators installed the walls and ceilings and placed insulation, soundproofing, and firestopping materials behind and onto those walls and ceilings. They may also have applied textures and trims to enhance both the interiors and exteriors of the buildings.

There were approximately 150,000 drywall applicators working in the United States in 2004. By 2014, the number is expected to grow by 10,000. Depending on the level of experience and region of the country, annual wages range from $22,000 to $60,000, with most earning $28,000 to $36,000. Careers can progress from installer to specialty finisher to business owner, and related professions include sheetrock applicator and acoustical carpenter.

This is the second installment of a two-level curriculum that meets the requirements of a standard drywall applicator apprenticeship program (a minimum of 288 hours of instruction and 4,000 hours of on-the-job training). In the first level of drywall training, you learned the basics, including insulation, installation, and finishing. In the second level, you'll explore specialized topics, such as steel framing, acoustical ceilings, and specialty finishes. In both levels, you'll learn about the materials and tools used in the drywall profession.

This curriculum was developed by a panel of subject matter experts from the drywall industry and produced by NCCER. By choosing to continue drywall training, you are taking a step toward a satisfying and rewarding career.

We wish you success as you begin your construction career in the drywall trade, and hope that you'll continue your training outside of this series. Continuing your craft education is important, as technology and the materials with which you'll work are changing all the time. By taking advantage of training opportunities as they arise, you'll demonstrate initiative and a desire to learn—qualities that are present in the industry's best professionals.

We also invite you to visit the NCCER website at **www.nccer.org** for the latest releases, training information, the *Cornerstone* magazine, and much more. You can also reference the Pearson product catalog online at **www. crafttraining.com**. Your feedback is welcome. You may email your comments to **curriculum@nccer.org** or send general comments and inquiries to **info@nccer.org**.

NCCER STANDARDIZED CURRICULA

NCCER is a not-for-profit 501(c)(3) education foundation established in 1996 by the world's largest and most progressive construction companies and national construction associations. It was founded to address the severe workforce shortage facing the industry and to develop a standardized training process and curricula. Today, NCCER is supported by hundreds of leading construction and maintenance companies, manufacturers, and national associations. The NCCER Standardized Curricula was developed by NCCER in partnership with Pearson Education, Inc., the world's largest educational publisher.

Some features of NCCER's Standardized Curricula are as follows:

- An industry-proven record of success
- Curricula developed by the industry for the industry
- National standardization providing portability of learned job skills and educational credits
- Compliance with the Office of Apprenticeship requirements for related classroom training (*CFR* 29:29)
- Well-illustrated, up-to-date, and practical information

NCCER also maintains a Registry that provides transcripts, certificates, and wallet cards to individuals who have successfully completed modules of NCCER's Standardized Curricula. *Training programs must be delivered by an NCCER Accredited Training Sponsor in order to receive these credentials.*

Contents

NCCER Standardized Curricula

NCCER's training programs comprise over 80 construction, maintenance, and pipeline areas and include skills assessments, safety training, and management education.

Boilermaking
Cabinetmaking
Carpentry
Concrete Finishing
Construction Craft Laborer
Construction Technology
Core Curriculum: Introductory Craft Skills
Drywall
Electrical
Electronic Systems Technician
Heating, Ventilating, and Air Conditioning
Heavy Equipment Operations
Highway/Heavy Construction
Hydroblasting
Industrial Coating and Lining Application Specialist
Industrial Maintenance Electrical and Instrumentation Technician
Industrial Maintenance Mechanic
Instrumentation
Insulating
Ironworking
Masonry
Millwright
Mobile Crane Operations
Painting
Painting, Industrial
Pipefitting
Pipelayer
Plumbing
Reinforcing Ironwork
Rigging
Scaffolding
Sheet Metal
Signal Person
Site Layout
Sprinkler Fitting
Tower Crane Operator
Welding

MARITIME

Maritime Industry Fundamentals
Maritime Pipefitting
Maritime Structural Fitter

GREEN/SUSTAINABLE CONSTRUCTION

Building Auditor
Fundamentals of Weatherization
Introduction to Weatherization
Sustainable Construction Supervisor
Weatherization Crew Chief
Weatherization Technician
Your Role in the Green Environment

ENERGY

Alternative Energy
Introduction to the Power Industry
Introduction to Solar Photovoltaics
Introduction to Wind Energy
Power Industry Fundamentals
Power Generation Maintenance Electrician
Power Generation I&C Maintenance Technician
Power Generation Maintenance Mechanic
Power Line Worker
Power Line Worker: Distribution
Power Line Worker: Substation
Power Line Worker: Transmission
Solar Photovoltaic Systems Installer
Wind Turbine Maintenance Technician

PIPELINE

Control Center Operations, Liquid
Corrosion Control
Electrical and Instrumentation
Field Operations, Liquid
Field Operations, Gas
Maintenance
Mechanical

SAFETY

Field Safety
Safety Orientation
Safety Technology

SUPPLEMENTAL TITLES

Applied Construction Math
Tools for Success

MANAGEMENT

Fundamentals of Crew Leadership
Project Management
Project Supervision

SPANISH TITLES

Acabado de concreto: nivel uno
Aislamiento: nivel uno
Albañilería: nivel uno
Andamios
Carpintería: Formas para carpintería, nivel tres
Currículo básico: habilidades introductorias del oficio
Electricidad: nivel uno
Herrería: nivel uno
Herrería de refuerzo: nivel uno
Instalación de rociadores: nivel uno
Instalación de tuberías: nivel uno
Instrumentación: nivel uno, nivel dos, nivel tres, nivel cuatro
Orientación de seguridad
Paneles de yeso: nivel uno
Seguridad de campo

Acknowledgments

This curriculum was revised as a result of the farsightedness and leadership of the following sponsors:

Associated Builders and Contractors, New Mexico Chapter

Associated Builders and Contractors, Pennsylvania Chapter

Baker Drywall

C.J. Coakley Co., Inc.

Greenville Technical College

Marek Brothers Systems, Inc.

This curriculum would not exist were it not for the dedication and unselfish energy of those volunteers who served on the Authoring Team. A sincere thanks is extended to the following:

David Ackerman

Bruce Bostick

Bob Consroe

Arthur Ehmling

Harry Parker

Robert Pelletier

Noe Vela

NCCER PARTNERS

American Fire Sprinkler Association

Associated Builders and Contractors, Inc.

Associated General Contractors of America

Association for Career and Technical Education

Association for Skilled and Technical Sciences

Construction Industry Institute

Construction Users Roundtable

Construction Workforce Development Center

Design Build Institute of America

GSSC – Gulf States Shipbuilders Consortium

ISN

Manufacturing Institute

Mason Contractors Association of America

Merit Contractors Association of Canada

NACE International

National Association of Minority Contractors

National Association of Women in Construction

National Insulation Association

National Technical Honor Society

National Utility Contractors Association

NAWIC Education Foundation

North American Crane Bureau

North American Technician Excellence

Pearson

Pearson Qualifications International

Prov

SkillsUSA®

Steel Erectors Association of America

U.S. Army Corps of Engineers

University of Florida, M. E. Rinker School of Building Construction

Women Construction Owners & Executives, USA

45201-09

Commercial Drawings

45201-09
Commercial Drawings

Topics to be presented in this module include:

Overview

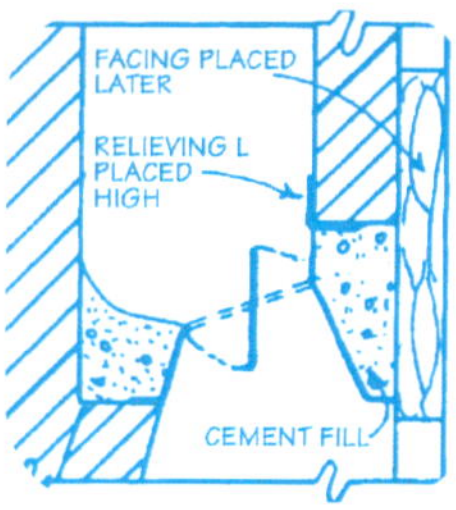

To build commercial structures, you must be able to understand and interpret plans and drawings. The project drawings will tell you everything from where to place the building on the site to how to construct an exterior wall. The specifications provide details on materials and construction methods. The basic principles such as scaling and dimensioning apply to both residential and commercial drawings. However, commercial drawings are much more detailed, and can contain additional structural, mechanical, and electrical systems. They require close study and review. Understanding the plan will help keep construction activities well-organized and true to the design.

Objectives

When you have completed this module, you will be able to do the following:

1. Recognize the differences between commercial and residential construction drawings.
2. Identify the sheet index, material symbols, code summary, project team, and other references contained in a set of commercial drawings.
3. Accurately read a set of commercial drawings.
4. Identify and document specific items from a door and window schedule.
5. Explain basic construction details and concepts employed in commercial construction.

Trade Terms

ADA requirements
As-builts
Beams
Building codes
Callouts
Civil drawings
Compressive strength
Contour lines
Elevation view
Girders
Isometric drawing
Joists
Landscape drawings
Plan view
Riser diagram
Sections
Schedules
Tensile strength
Title block

Prerequisites

Before you begin this module, it is recommended that you successfully complete *Core Curriculum: Introductory Craft Skills* and *Drywall Level One*.

This course map shows all of the modules in the second level of the *Drywall* curriculum. The suggested training order begins at the bottom and proceeds up. Skill levels increase as you advance on the course map. The local Training Program Sponsor may adjust the training order.

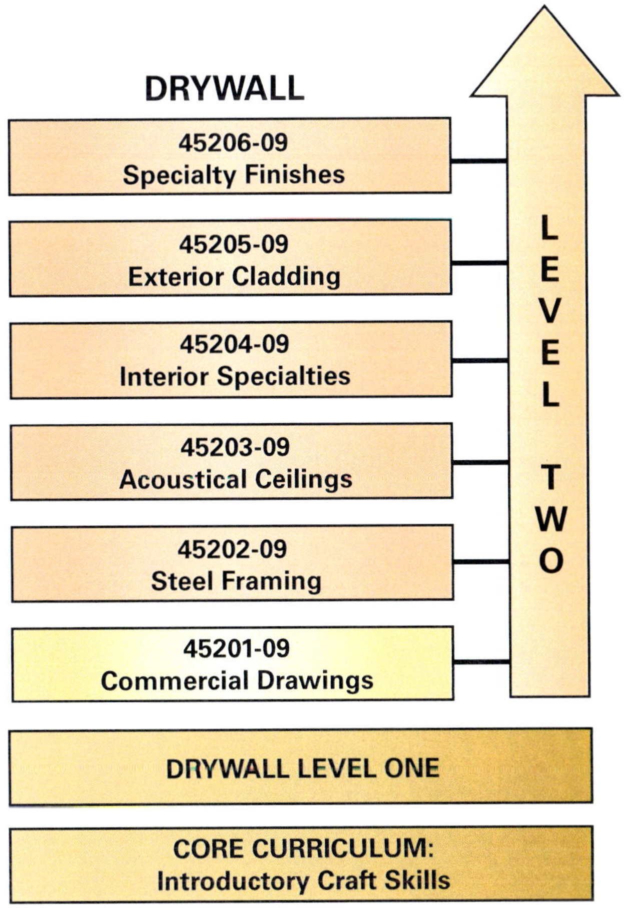

201CMAP.EPS

1.0.0 ♦ INTRODUCTION

A large part of the construction market in the United States is commercial and industrial construction. Commercial buildings come in various shapes and sizes (*Figure 1*). To build commercial structures, you must be able to understand and interpret the architect's plans and drawings. Commercial plans and drawings are usually more complex than residential plans. This module will introduce you to commercial drawings. The basic principles, such as scaling and dimensioning, are the same as those that apply to residential drawings. However, commercial structures have more dimensions to interpret. Additional section and detail drawings mean there will be more changes in scale.

2.0.0 ♦ REQUIREMENTS AND CONTENTS

Commercial or industrial construction work is often called heavy construction. This is because heavy equipment and heavy materials are used for these jobs. Cranes, hoists, and graders are heavy equipment. Heavy materials include steel and concrete. Most commercial projects are larger and more complicated than residential projects. They require a greater variety of construction techniques, equipment, and materials.

Consequently, plans and drawings for a commercial project are also more complicated than residential plans and drawings. Commercial structures have many different uses. Safety and environmental requirements must also be considered. The complexity of the commercial project is reflected in the complexity of the plans.

2.1.0 Requirements for Commercial Plans

There are several reasons why commercial construction plans are more detailed than residential plans. First, the structures are usually larger and more expensive to build. Another major consideration is legal liability. The contractor's legal liability is far greater in commercial construction because there are more applicable codes, ordinances, and regulations. Other reasons why commercial plans are more complex include the following:

- The architectural plans, drawings, **schedules**, and specifications are legal documents. Many state, local, and federal agencies demand greater detail in commercial construction drawings to substantiate any legal disputes.

201F01.EPS

Figure 1 ♦ Examples of modern commercial buildings.

- Code restrictions and safety requirements for commercial and industrial buildings are far more complicated than those for residential construction. More detailed drawings are used to make certain that all codes and local ordinances are met.
- The size of a commercial building requires a greater number of drawings, **sections**, details, and schedules, with more detail required to correlate the various parts of the structure.
- The materials used in commercial construction call for more detailed information on construction techniques, especially for structural steel.

A major commercial project may have hundreds of drawings in its plan set, plus associated schedules. Specialty subcontractors will generally get a partial set of the plans relating to their work. However, they should request a full set, including the room finish schedules. At least one complete set of plans will be kept in the field office for reference.

Brick Upon Brick

GOING GREEN

For nearly 40 years, the Empire State Building was the tallest building in the world. Approximately 10 million bricks were used in its construction. It was completed in 1931. Today it is likely that curtain walls containing brick facades would be used in place of individual bricks.

Despite the enormity of the project, construction of the Empire State Building was completed in about 15 months. One of the methods used to speed up construction was to have trucks unload the bricks down a chute, instead of dumping them in the street. The chute led to a large hopper from which the bricks were then loaded into carts, and hoisted to the location where they were needed. This innovative technique eliminated the back-breaking work of moving bricks from the brick pile to the bricklayer using a wheelbarrow.

Even though the Empire State Building was built in the 1930s, it has undergone extensive upgrades and is a charter member of the Energy Star Building Label Program. Energy Star is a joint program of the U.S. Environmental Protection Agency and the U.S. Department of Energy that helps to protect the environment by using energy-efficient products and practices.

201SA01.EPS

During the course of a project, **as-builts** are created using a set of working drawings, sometimes called record drawings, that incorporate (through revisions or annotations) any changes of dimensions, materials, form, and method of construction encountered in the completion of the structure or site. All changes, no matter how minor, as well as all clarifications are entered into a change database and recorded by the general contractor on two sets of redlined working or record drawings. Changes and clarifications refer to shop drawings, field engineering change proposals (ECPs), change orders, modifications, and requests for information (RFIs) which have been reviewed and approved by the project engineers. Once the project is complete, the as-builts are normally formalized into a permanent set of as-built drawings by the architectural firm and copies are distributed to the project owners as well as other designated parties.

2.2.0 Commercial Plan Contents

Construction drawings consist of several different kinds of drawings assembled into a set (*Figure 2*). Each type of drawing is assigned a letter. For each type, there may be several drawings, which are then numbered. For example, the first few electrical drawings would be numbered E1, E2, and E3. A complete set of commercial construction plans typically includes the following drawing types:

- Architectural – A
- Structural – S
- Mechanical – M
- Plumbing – P
- Electrical – E

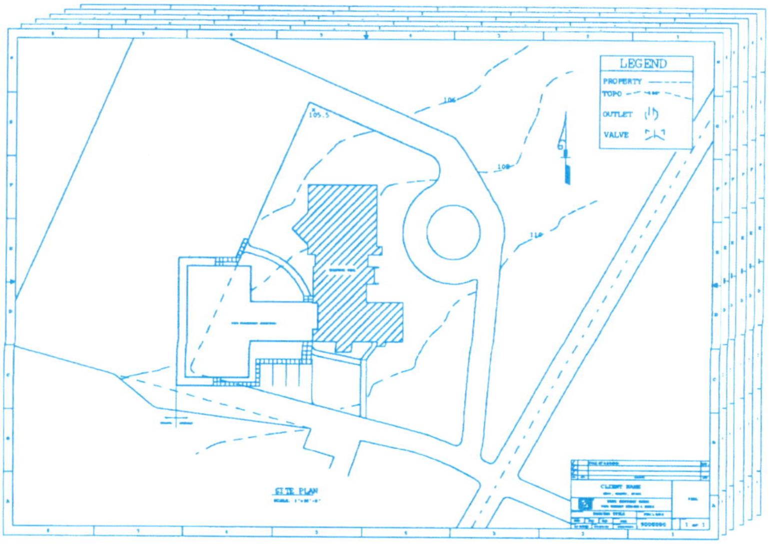

TITLE SHEET(S)

ARCHITECTURAL DRAWINGS

- SITE (PILOT) PLAN
- FLOOR PLANS
- INTERIOR/EXTERIOR ELEVATIONS
- SECTIONS
- DETAILS
- SCHEDULES

STRUCTURAL DRAWINGS

- FOUNDATION PLAN

PLUMBING DRAWINGS

MECHANICAL DRAWINGS

ELECTRICAL DRAWINGS

201F02.EPS

Figure 2 ◆ Drawing set.

The exact content of the plan set will vary, depending on the type and size of the job and the local code requirements. For example, a drawing set for a commercial office building may include **landscape drawings**. These would be identified by the letter L. In some commercial drawing sets the site plans are called **civil drawings**. They are marked C1, C2, C3, and so forth. Specifications and schedules may be referenced or included with each type of drawing. These two components are similar in content, if not in quantity, to architectural drawings for residences. *Figure 3* shows common drawing lines, and *Figure 4* shows standard symbols for various materials.

Object lines show the main outline of the structure. Dimension lines and extension lines show the size of an object. Leader lines are used to connect the notes to features on the drawing. Arrows indicate the extent of the measurement. The measurement is written above the dimension line or in a break.

Section lines, or cutting plane lines, show the location of a section view of that particular part of the structure. Designers use various formats to indicate sections. Usually, they are in the form of a U with arrows at the tips. A letter identifies the section. The section drawing is marked with a corresponding letter. The section view is placed either on the same page as the main drawing or all section views are placed on one sheet with each view referenced by the section letter.

Symbols provide another graphic indication for different types of materials. There are many commonly used symbols. However, there are no standardized symbols for specific materials. The best known reference for standard symbols, lines, and abbreviations is *Architectural Graphic Standards* prepared by the American Institute of Architects. Large commercial projects will have a key block or legend. It may be on the index page or a separate page. It lists the symbols and abbreviations used throughout the plan set.

INSIDE TRACK

Safeguarding the Drawing Set

Always treat a drawing set with care. It is best to keep two sets, one for the office and one for field use. Never remove original drawings or sheets from the set of original drawings in the office. Removal of original drawings from the office may result in their loss or damage. It can also cause delays and lost time for others who need to use or reproduce the drawings. Make sure that both sets are updated with any as-builts and other revisions or change orders. Pay attention to the revision date and applicable color codes to ensure that the correct drawing set is being used.

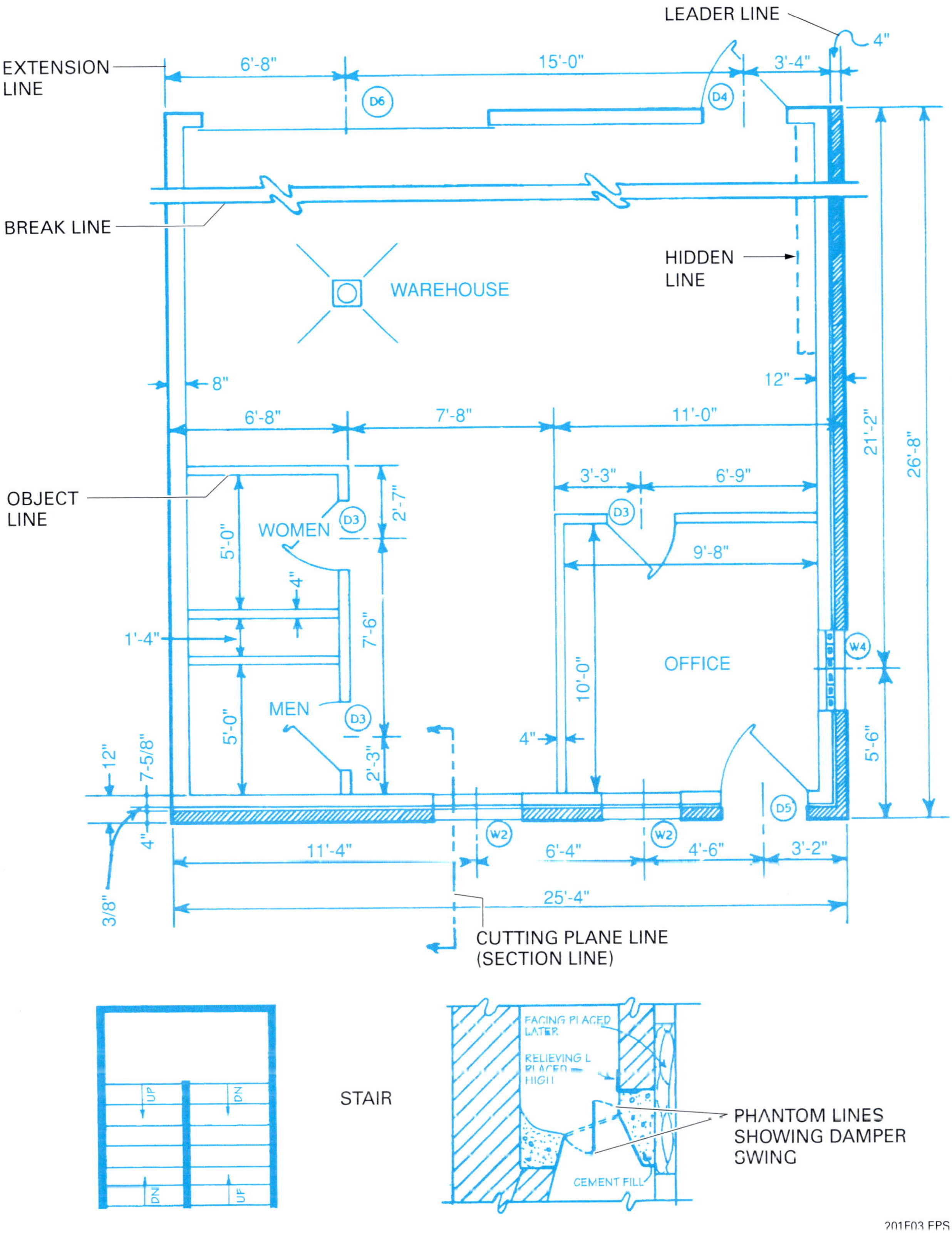

Figure 3 ◆ Drawing lines.

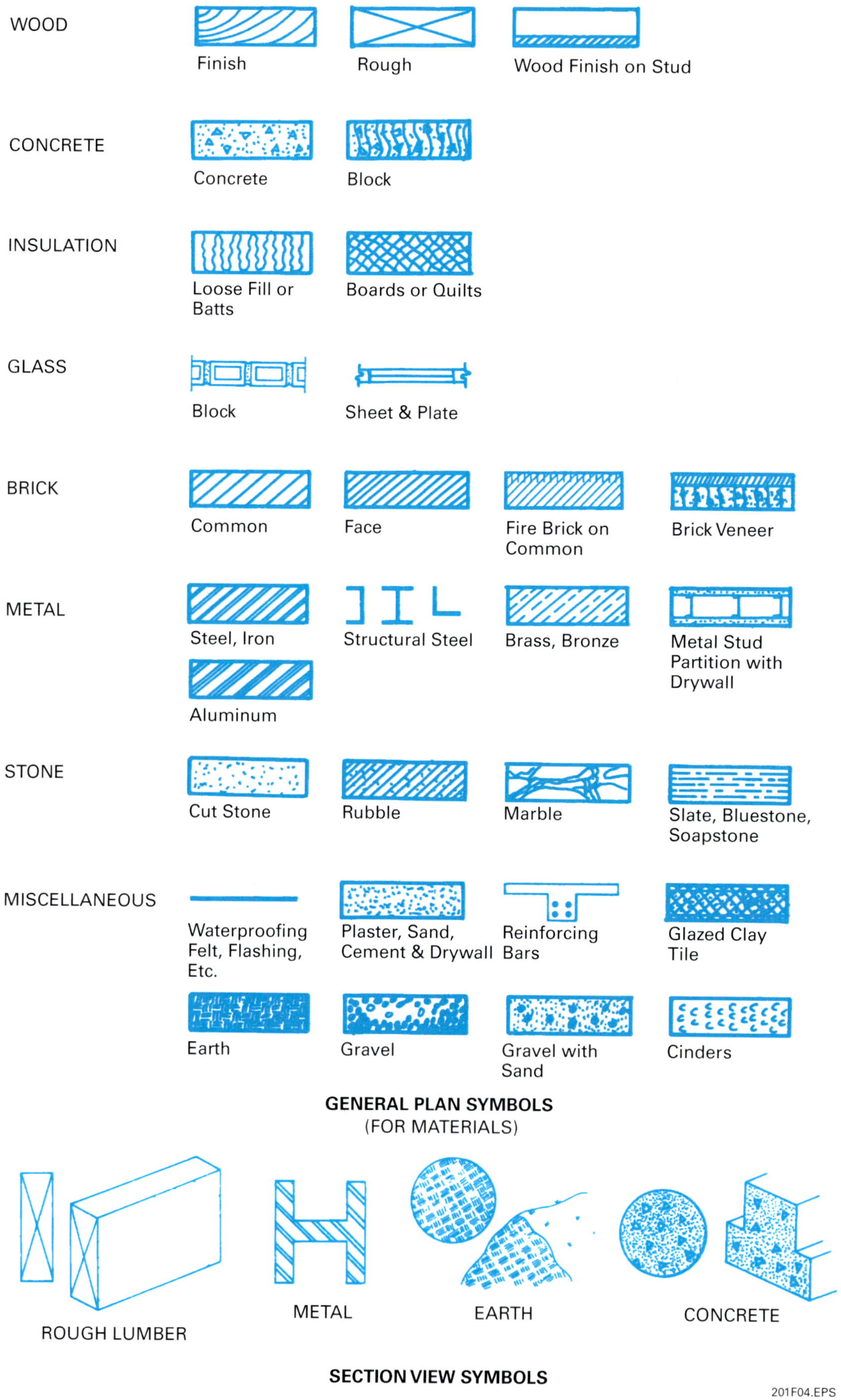

Figure 4 ◆ Material symbols.

2.2.1 Architectural Drawings

The architectural drawings are usually labeled with page numbers beginning with the letter A. These contain general design features of the building, room layouts, construction details, and materials requirements. The architectural drawings will include the following:

- A site plan and/or other location plans
- Floor plans
- Wall sections
- Reflected ceiling plan
- Door and window details and schedules
- Partition details (*Figure 5*)
- Elevations
- Special application details including finish details and schedules

The site plan shows topographic features including trees, bodies of water, and ground cover. It will also show man-made features such as roads, railroad tracks, and utility lines. Typical topographic symbols are shown in *Figure 6*.

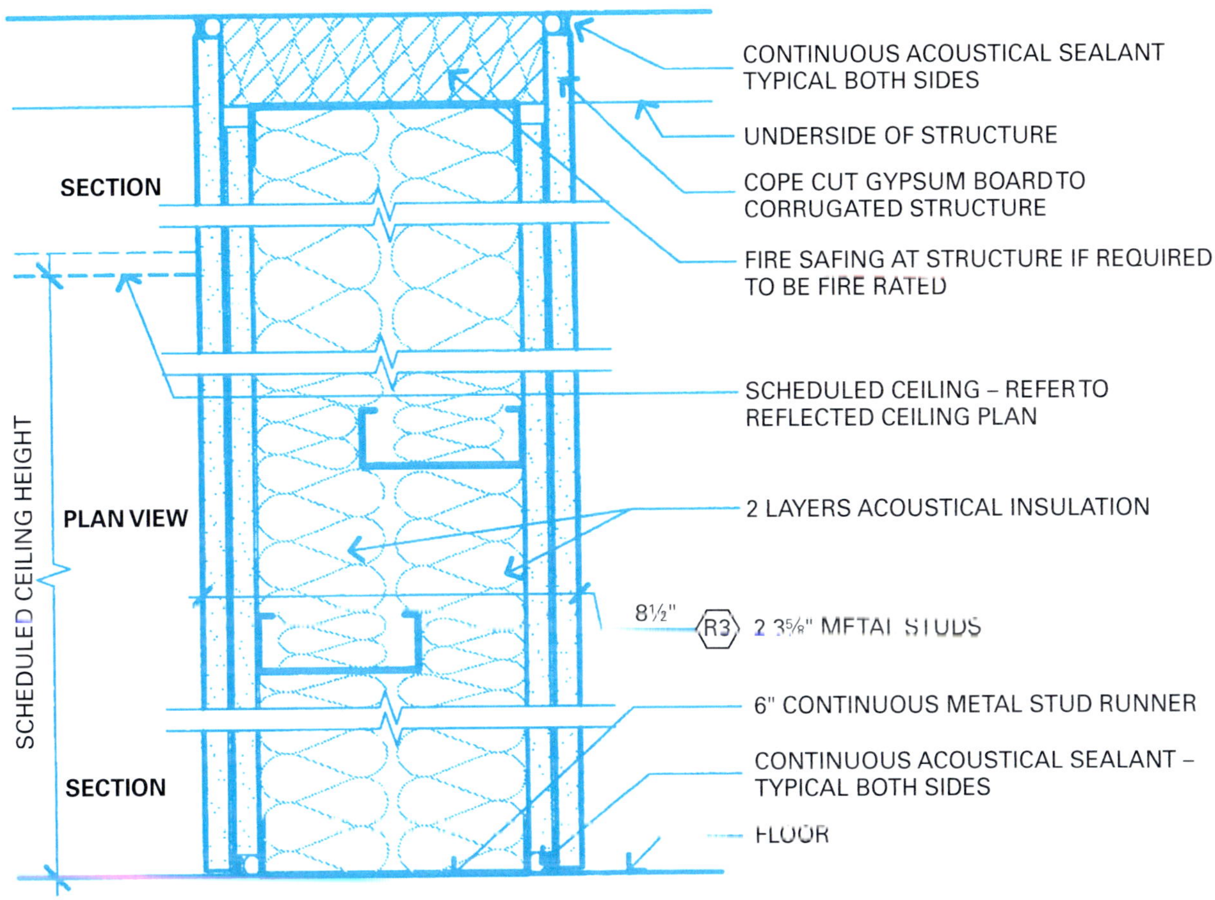

25 GAUGE METAL STUDS AT 12" O.C. BUT STAGGERED WITH 2 LAYERS 3" ACOUSTICAL INSULATION WOVEN BETWEEN THE STUDS AND TWO LAYERS OF GYPSUM BOARD EACH SIDE.

PARTITION TYPE	L/240 LIMITING HEIGHT	FIRE RATING	STC SOUND TRANSMISSION CLASS
R3	13'6"	2 HOUR – UL U411	60

Partition Type | **R**

Scale: 3"= 1'-0"

201F05A.EPS

Figure 5 ◆ Typical partition details (1 of 2).

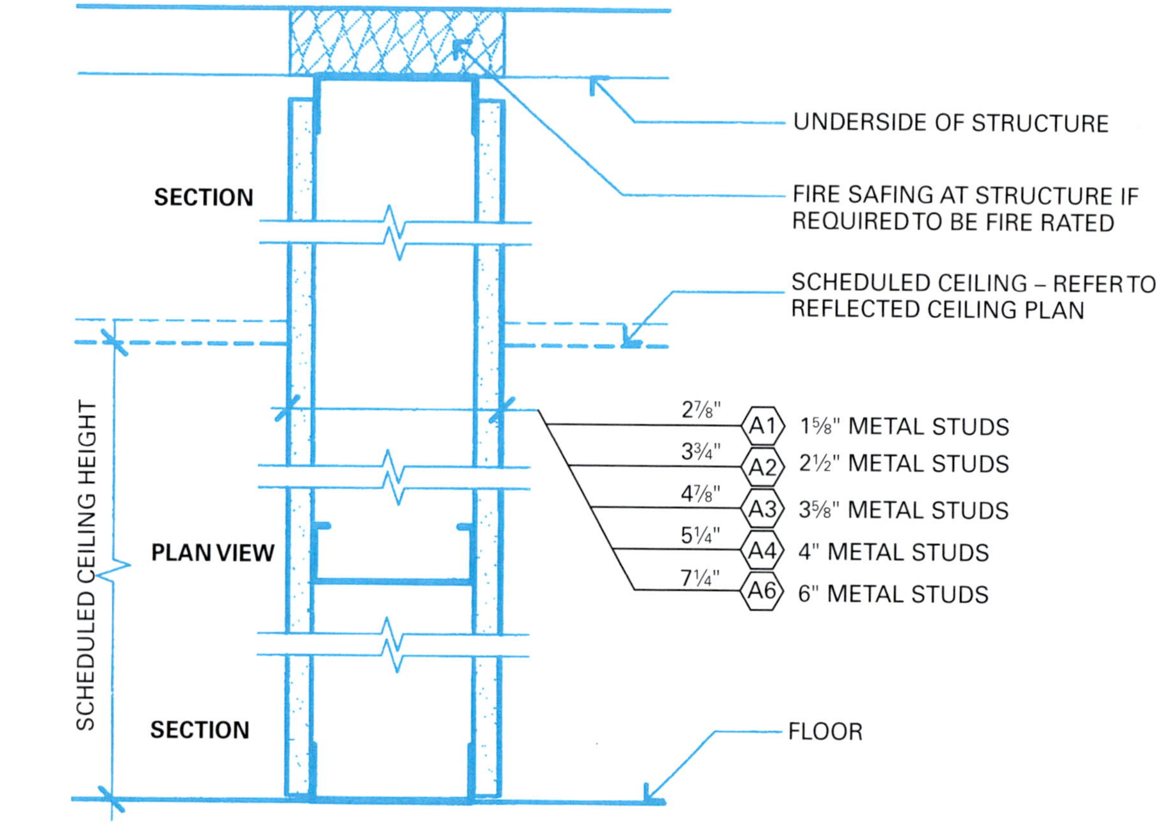

25 GAUGE METAL STUDS WITH ONE LAYER OF ⅝" GYPSUM BOARD EACH SIDE

PARTITION TYPE	L/240 LIMITING HEIGHT	FIRE RATING	STC SOUND TRANSMISSION CLASS
A1	9'6"	–	–
A2	12'6"	1 HOUR – UL U494	40
A3	16'	1 HOUR – UL U465	40
A4	17'3"	1 HOUR – UL U465	40
A6	20'	1 HOUR – UL U465	40

Partition Type A
Scale: 3"= 1'-0"

201F05B.EPS

Figure 5 ◆ Typical partition details (2 of 2).

2.2.2 Structural Drawings

Structural drawings provide a view of the structural members of the building and how they will support and transmit those loads to the ground. Structural drawings are numbered sequentially and designated by the letter S. They are normally located after the architectural drawings in a plan set.

A structural engineer prepares the structural drawings. They must calculate the forces on the building and the load that each structural member must withstand. The structural support information includes the foundation, size, and reinforcing requirements; the structural frame type and size of each member; and details on all connections required. The structural drawings usually include the following:

- Foundation plans
- Structural framing plans for floors and roofing
- Structural support details
- Notes to describe construction and code requirements

Structural support for a commercial building may be steel framing, precast concrete structural elements, or cast-in-place concrete. Unlike residences, modern commercial buildings do not have wooden frames, except in unusual circumstances.

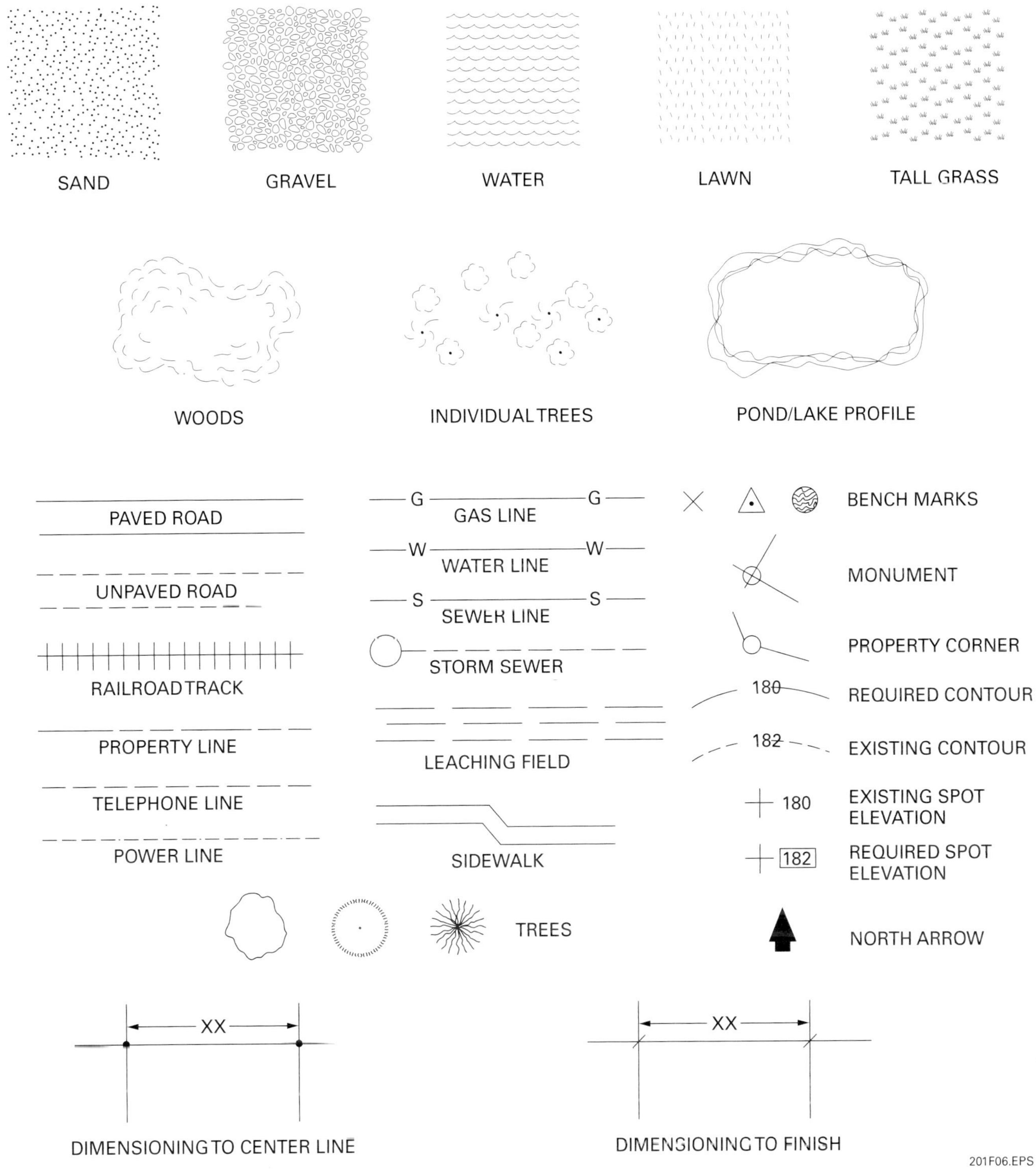

Figure 6 ♦ Topographic symbols.

Structural drawings provide useful information. They can stand alone for craftworkers such as framers and erectors. The structural drawings show the main building members and how they relate to the interior and exterior finishes. They do not include information that is unnecessary at the structural stage of construction.

Structural drawings start with the foundation plans. Foundation plans are followed by ground-floor or first-floor plans, upper-floor plans, and the roof plan. Only information essential to the structural systems is shown. For example, a second-floor structural plan would show the steel or concrete framing and the configuration and spacing of the loadbearing members. The walls, ceilings, or floors would not be shown.

2.2.3 Mechanical Drawings

Mechanical drawings show the different mechanical systems of the building. Specifically, they include the HVAC, plumbing, and fire protection drawings. These drawings have a prefix of M, P, and FP respectively. The **plan view** format is commonly used on these drawings. This view offers the best illustration of the location and configuration of the work. The drawings serve as a diagram of the system layout.

A large amount of information is required for mechanical work. There is limited space on the drawing to show the piping, valves, and connections. Special symbols are used for clarity. *Figure 7* contains some common HVAC symbols.

Detail drawings are sometimes used on mechanical drawings. Unlike the details shown on architectural drawings, these detail drawings are not normally drawn to scale. Usually, they are drawn as an elevation or perspective view. They show details about the configuration of the equipment.

2.2.4 Plumbing Drawings

Plumbing drawings are considered part of the mechanical plans. However, they are usually placed in their own set of drawings for clarity. Unless the building is very basic, placing them on the same sheets as the HVAC drawings would cause confusion. Plumbing drawings usually include the following:

- Site plan for water supply and sewage disposal systems
- Floor plans for the fire system, including hydrant connections and sprinkler systems
- Floor plans for the water supply system and fixture location
- Floor plans for the waste disposal system
- **Riser diagrams** to describe the vertical piping features
- Floor plans and riser diagrams for the gas lines

The most common symbols used to designate the water and gas systems on plumbing drawings are shown in *Figure 8*. In addition to these symbols, there will be specific graphic symbols for the layout of such items as toilets, sinks, water heaters, and sump pumps. These will vary from structure to structure, depending on the specific design.

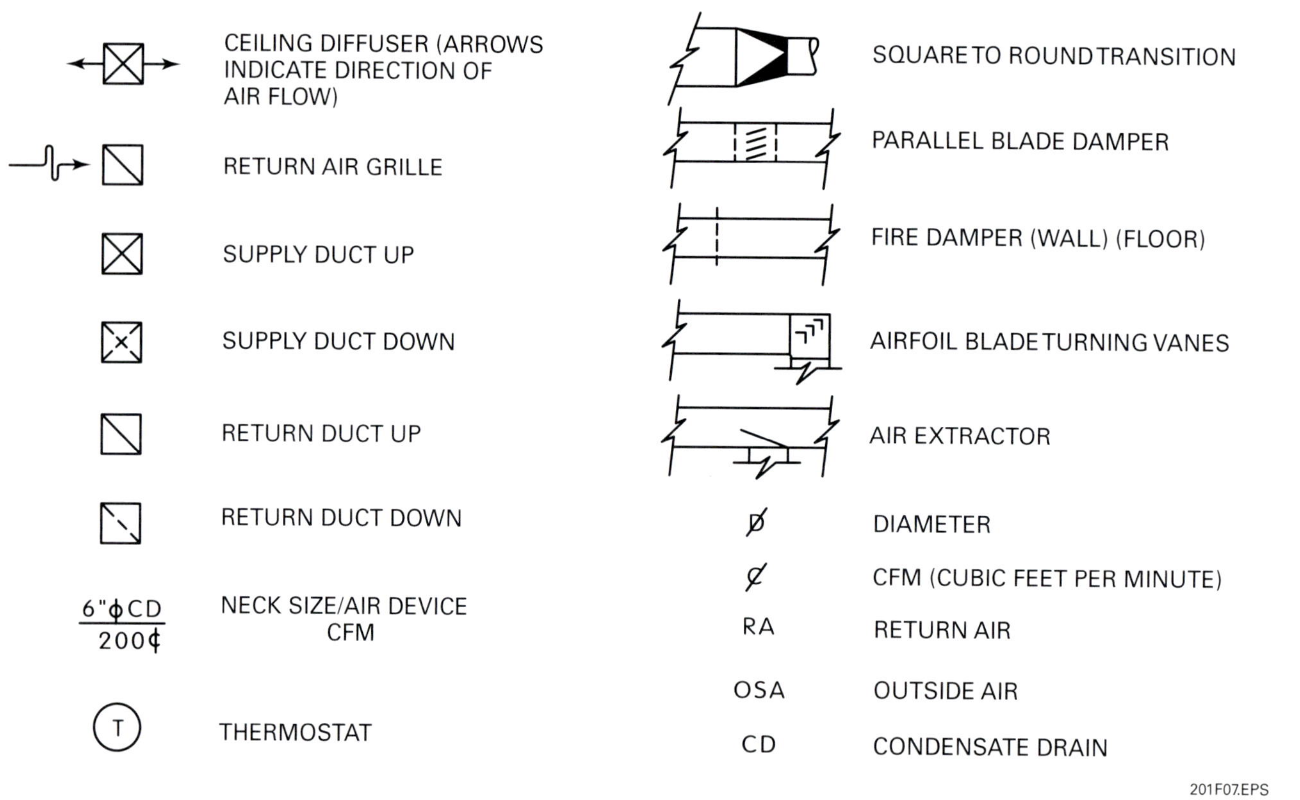

Figure 7 ♦ HVAC symbols.

Figure 8 ◆ Plumbing symbols.

2.2.5 Electrical Drawings

The last part of the plan set usually contains the electrical drawings. They show the various electrical and communications systems of the building.

The electrical drawings are labeled with page numbers beginning with the letter E. They contain information on the electrical service requirements for the building. They also show the location of all outlets, switches, and fixtures. In addition to schematics of the branch circuits for the building, electrical drawings will include the following:

- Site plan for electrical service requirements
- Floor plans for the outlet and switch locations and the branch circuit requirements
- Lighting plans

- Emergency power and lighting systems
- Life safety systems
- Any backup power generation facilities
- Notes and details to describe other parts of the electrical system

Like mechanical drawings, electrical drawings use the plan view to show system layout. Details and schedules provide clarification. One drawing may include power, lighting, and telecommunications layouts. In more complex structures, the systems are shown separately. There are many different symbols for electrical connections and fixtures. Commonly used symbols are shown in *Figure 9*. Special symbols for components such as power supplies, security systems, and circuit boards are usually designated by the manufacturer.

2.2.6 ADA (Americans with Disabilities Act) Requirements

All trades should be aware of the **ADA requirements** for other trades and coordinate whenever possible. For example, if you are finishing a bathroom and note that the plumbers put the toilet trap at the wrong floor spacing, it is best to note this before finishing and avoid a callback. Common ADA requirements include such things as doorway width and hardware height, plumbing fixture height, electrical switch and receptacle height, thermostat location, and so on.

3.0.0 ◆ READING AND UNDERSTANDING DRAWINGS

When reading commercial plans, it is best to follow a step-by-step process to avoid confusion and to catch all the important details. Use the following list as a guide:

Step 1 Begin by reading the project specifications to pick up details not found on the drawings. Remember that in cases of conflicting information, the specifications take precedence over the drawings, and the structural drawings take precedence over the architectural drawings.

Step 2 Quickly review all drawings that give a general impression of the shape, size, and appearance of the structure, including the site plan, floor plans, and exterior **elevation views**.

Step 3 Begin correlating the floor plans with the exterior elevations, making certain that the parts appear to fit together logically.

Step 4 Next, look at the wall sections and determine the wall types (materials, load-bearing or nonbearing) and construction details or procedures.

Step 5 Review the structural plans. Determine the foundation requirements and the type of structural system. Turn back to the site plan, floor plans, wall sections, and elevations as often as needed to correlate the structural and architectural drawings.

Step 6 Review all details on the architectural and structural drawings. Carefully consider all items that may require special construction procedures. It may be necessary to label detail drawings with related floor plan and elevation information. This is known as creating a reverse trail from the detail drawings back to the larger view.

Step 7 Review all interior elevations and try to get a clear picture of what the interior of the building will look like.

Step 8 Review the finish schedule.

Step 9 Review the mechanical and electrical plans. Coordinate with other trades to ensure proper sequencing.

If any part of a drawing is not clear, check with your supervisor. There may be a logical explanation for the item in question, or the plans may require clarification. An RFI is generated to resolve these conflicts. The architect/engineer should clarify and resolve all conflicts in writing so there will not be any confusion later in the project.

INSIDE TRACK

Drawing Revisions

When a set of drawings has been revised, always make certain that the most up-to-date set is used for all future work. Either destroy the old, obsolete drawing or else clearly mark on the affected sheets: Obsolete Drawing – Do Not Use. A good practice is to remove the obsolete drawings from the set and file them as history copies for future reference.

Also, when working with a set of construction drawings and written specifications for the first time, thoroughly check each page to see if any revisions or modifications have been made to the original. Doing so can save time and expense for all concerned.

Figure 9 ◆ Electrical symbols.

3.1.0 Architectural Drawings

Architectural drawings are the core drawings of any plan set. They are sequentially numbered, usually starting with the site plan or the basement. In some cases, the exterior drawings, such as the site plan and landscaping plans, are numbered separately from the architectural drawings. If this is the case, the architectural drawings are sequentially numbered in order of basement or ground-floor plans; upper-level floor plans; exterior elevations; sections; interior elevations; details; and window, door, and room finish schedules.

3.1.1 Site Plans

The main purpose of the site plan is to locate the structure within the confines of the building lot. The site plan clearly shows the building's dimensions (*Figure 10*). The building is usually shown by the size of the foundation and the distances to the respective property lines.

A commercial building needs a site that is appealing, convenient for customer traffic, and in harmony with other commercial structures in the area. Commercial site drawings show many of the same features as residential drawings, but are

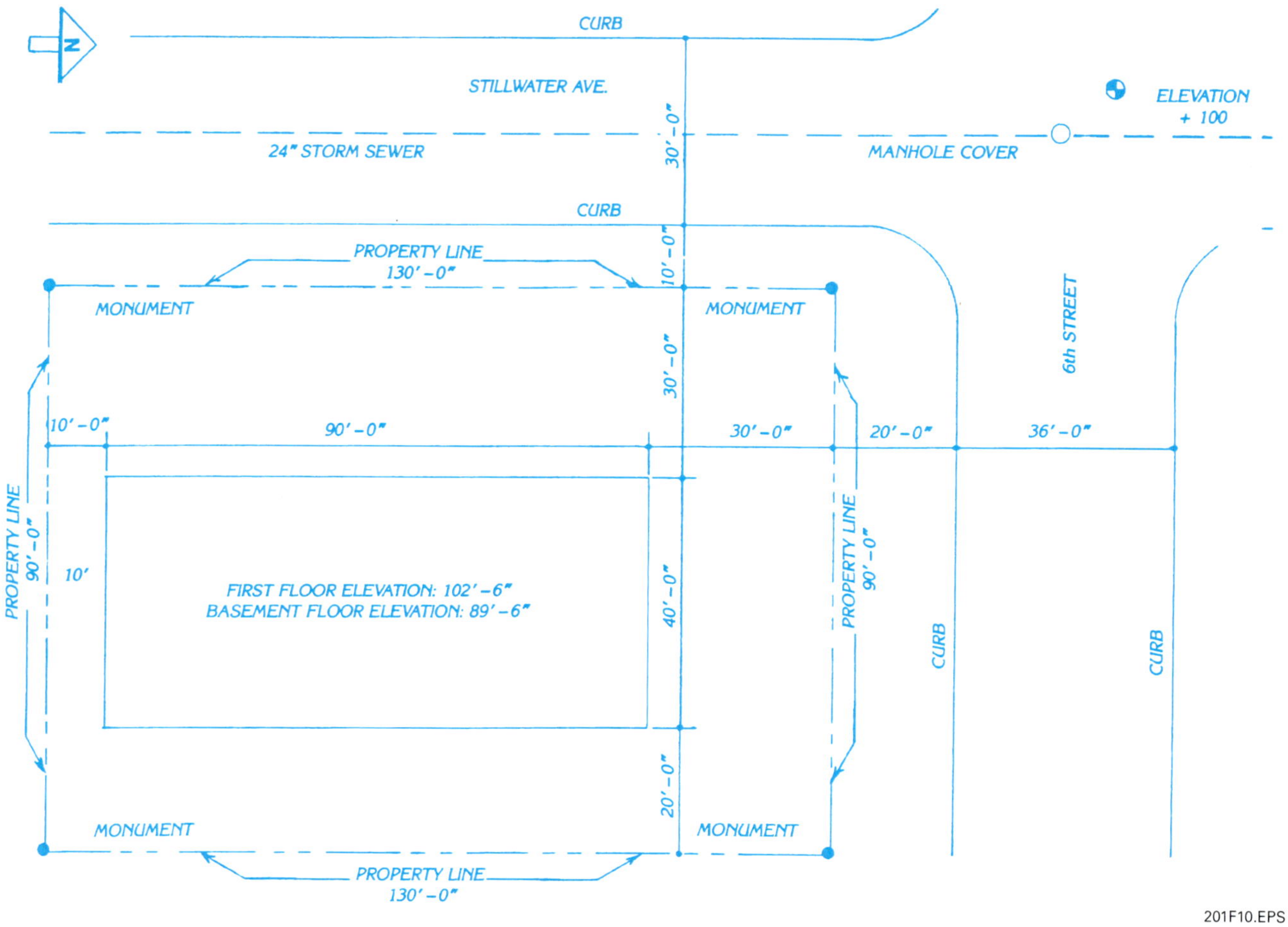

Figure 10 ◆ Site plan.

far more detailed. They include details on site improvement features, existing and finish **contour lines**, paving areas, and site access. Since a commercial structure serves the public, the architect, owner, and zoning board should note and specify all site changes. Commercial site plans will also show the location of public utilities.

When looking at a commercial site plan for the first time, notice the features that also appear on residential location plot plans, including the following:

- Survey data such as directional arrow, property lines, and the structure's relationship to other features
- Geographic data such as lot corner elevations, existing and proposed contours, and landscaping features
- Building features such as floor elevations, exterior wall positioning, roof overhang, and drainage

Features on the commercial site plans that may not be found on residential plans include the following:

- Details of paved areas, including walkways, driveways and parking areas
- Details on grading and other site work
- Notations which refer to details on the site plan or elsewhere in the drawings
- Position of existing and new utility lines
- Dimensions usually specified in feet and tenths of feet
- Referenced details, sections, and elevations which pertain to site preparation features
- Symbols or notations detailing materials types, sizes, and positions

Another primary purpose of the site plan is to show the unique surface conditions, or topography, of the lot. The topography of a particular lot may be shown right on the site plan. For projects in which the topography must be shown separately for clarity, a grading plan is used.

DID YOU KNOW?

Museum of American Architecture

The Athenaeum of Philadelphia is a museum of American architecture and interior design. The work of over 1,000 American architects from 1800 to 1945 is collected and available for research. The museum holds 150,000 drawings, 50,000 photographs, and many other documents. Most of the drawings are from Philadelphia, but the collection also includes drawings of buildings in most of the states and several countries. The building was designed in 1845 by John Notman. It was one of the first Philadelphia buildings built of brownstone.

201SA02.EPS

3.1.2 Floor Plans

Floor plans of both commercial and residential structures show various floor levels as if a horizontal plane had been cut through the structure. This results in an overhead view of each floor. One common use of the floor plan is to calculate the area of each room and other areas represented on the floor plan. This information is needed to determine the amount of floor covering, wallboard, and other material required. For most rooms, it is simply a matter of multiplying the length and width dimensions to determine the square footage of the room. If the room is not a perfect rectangle, however, the calculation must be done by breaking the room into pieces, then adding the square footage of the pieces.

The most noticeable difference between commercial and residential floor plans is the amount of detail. Commercial plans contain more details of room use, finishes, wall types, sound transmission, and fire retardation. In fact, most commercial floor plans will incorporate a legend or chart to specify the various interior wall types shown on the plan. The drawing scale is normally $\frac{1}{8}$ inch to 1 foot for the plan views. The detailing instructions usually include the following types of information:

- Revisions and changes that may be enclosed by a cloud outline
- Numerous **callouts** specifying sectional views and details
- Room assignment designations by function or number
- Detailed dimensioning of all visible parts of the structure
- Finish designations referenced to schedules
- Life safety drawings, which include requirements for such things as sprinkler systems, fire walls, and exit corridors

On commercial plans, little is left to chance for several reasons. First, the construction itself is varied and complex. Second, construction must meet all code specifications. Finally, different contractors will be working on the same job. Since there are many ways to accomplish the same task, the requirements are specified in detail to achieve consistency throughout the structure.

INSIDE TRACK: Color Codes

It is a good idea to color code different wall types on a floor plan using different colored markers. In addition to simplifying the installation, it will also ensure that you review the installation line by line, which provides an opportunity to plan the installation and anticipate any problems.

Each door or window on a floor plan for a commercial building is typically accompanied by a number, letter, or both. This number/letter is an identifier that refers to a door or window schedule that describes the corresponding door or window by size, type of materials, or model number for the specific door or window. Schedules are discussed in more detail later in this section.

When supplied, roof plans (*Figure 11*) provide information about the roof slope, roof drain placement, and other pertinent information. Where applicable, the roof plan may also show information on the location of air conditioning units, exhaust fans, and other ventilation equipment.

To help clarify ambiguous parts of the drawing, notes may be written on commercial floor plans. This is particularly true when any feature differs from one area to another. In most instances, these notes are important not only because they show variations, but also because they detail the responsibilities of those involved in the construction. A plan note may read: "Furnished by owner," "Refer to structural drawings," or "Not in contract." *Figure 11* includes a note instructing the contractor to coordinate the placement of gutters and downspouts with the architect. Be sure to read all notes, even if they apply to other trades. In addition, review all plans, including those of the other trades. They may contain information that pertains to your work.

For large buildings, the architect/engineer often divides the floor plan into sections by grid lines. The grid is the same for all floors of the building. Using the grid allows the architect and engineer to locate features very specifically anywhere in the building. The grid lines are

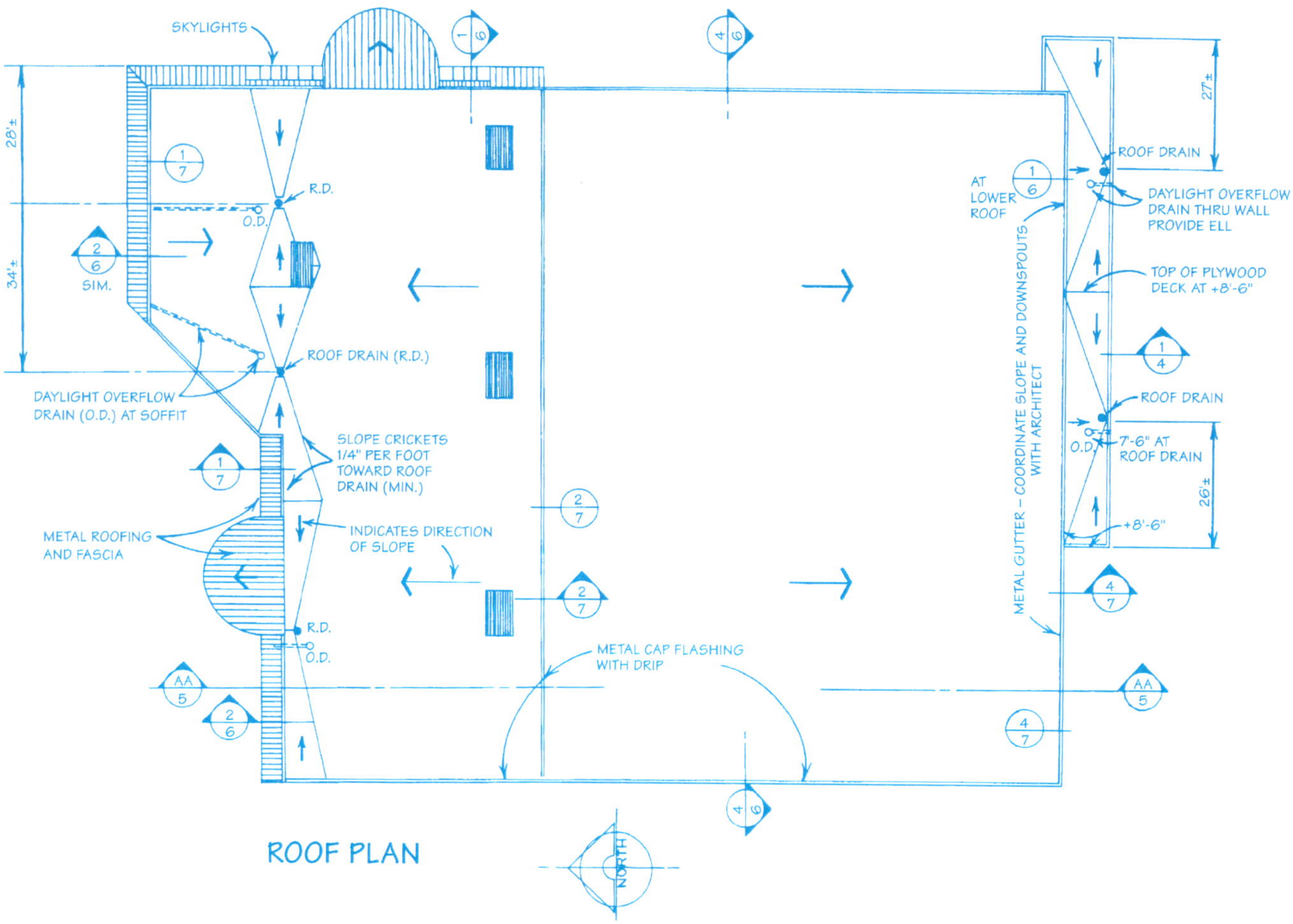

201F11.EPS

Figure 11 ◆ Roof plan.

useful for locating features that are repeated on one floor or from one floor to another. The grid markings on the floor plan also reappear on the structural drawings where they locate footings and columns.

3.1.3 Schedules and Details

Since a commercial building has so many varied parts and functions, it is impossible to draw all of the features on one drawing. Many of the smaller detail features are in notes on the actual drawings. Even then, the plans may become cluttered if too many notes appear. To prevent such a problem, schedules, detail drawings, and written specifications are used to describe the construction materials and procedures required.

In most cases, the architectural drawings will include schedules for doors, windows, and interior wall finishes (*Figure 12*). A schedule is actually a chart or table which provides detailed information corresponding to various parts of the

DOOR SCHEDULE

DOOR	DOOR SIZE (1 3/4" THICK UNLESS OTHERWISE NOTED)	DOOR		FIRE	FRAME			GLASS	HDW	REMARKS
		MAT'L	TYPE	RATING	ELEV.	TYPE	DETAIL		GRP.	
1011	3'-0" x 7'-0"	ALUM	A		6/A2.1	ALUM	22/A5.1		1	
1012	3'-0" x 7'-0"	ALUM	A		4/A2.1	ALUM	23/A5.1		1	
1031	3'-0" x 7'-0"	HM	B		HM-1	F1	2/A2.1		3	COORDINATE DOOR UNDERCUT WITH SPEC'D THRESHOLD
1041	3'-0" x 7'-0"	WOOD	C		HM-1	F1	1/A2.1	1/4" TEMP	4	
1051	3'-0" x 7'-0"	WOOD	C		HM-1	F1	1/A2.1	1/4" TEMP	4	
1052	3'-0" x 7'-0"	WOOD	C		HM-1	F1	1/A2.1	1/4" TEMP	4	
1071	3'-0" x 7'-0"	WOOD	C		HM-1	F1	1/A2.1	1/4" TEMP	5	
1081	3'-0" x 7'-0"	WOOD	C		HM-1	F1	1/A2.1	1/4" TEMP	5	
1091	3'-0" x 7'-0"	WOOD	B		HM-1	F2	1/A2.1		5	
1092	3'-0" x 7'-0"	HM	B		HM-1	F1	2/A2.1		2	COORDINATE DOOR UNDERCUT WITH SPEC'D THRESHOLD
1101	3'-0" x 7'-0"	WOOD	B		HM-1	F1	1/A2.1		6	
1111	3'-0" x 7'-0"	WOOD	B		HM-1	F1	1/A2.1		6	
1121	3'-0" x 7'-0"	HM	B	3 HR	HM-2	F1	3/A2.1		7	
1131	3'-0" x 7'-0"	HM	B		HM-1	F1	2/A2.1		2	COORDINATE DOOR UNDERCUT WITH SPEC'D THRESHOLD
1141	3'-0" x 7'-0"	WOOD	C		HM-1	F1	1/A2.1	1/4" TEMP	4	
1161	2'-4" x 7'-0" PAIR	WOOD	B		HM-1	F1	1/A2.1		8	4'-8" x 7' x 0" FRAME

WINDOW SCHEDULE

SYMBOL	WIDTH	HEIGHT	MAT'L	TYPE	SCREEN	QUANTITY	REMARKS	MANUFACTURER	CATALOG NUMBER
A	3'-8"	3'-0"	ALUM.	DOUBLE HUNG	YES	2	4 LIGHTS, 4 HIGH	LBJ Window Co.	141 PW
B	3'-8"	5'-0"	ALUM.	DOUBLE HUNG	YES	1	4 LIGHTS, 4 HIGH	LBJ Window Co.	145 PW
C	3'-0"	5'-0"	ALUM.	STATIONARY	STORM ONLY	2	SINGLE LIGHTS	H & J Glass Co.	59 PY
D	2'-0"	3'-0"	ALUM.	DOUBLE HUNG	YES	1	4 LIGHTS, 4 HIGH	LBJ Window Co.	142 PW
E	2'-0"	6'-0"	ALUM.	STATIONARY	STORM ONLY	2	20 LIGHTS	H & J Glass Co.	37 TS
F	3'-6"	5'-0"	ALUM.	DOUBLE HUNG	YES	1	16 LIGHTS, 4 HIGH	LBJ Window Co.	143 PW

201F12.EPS

Figure 12 ◆ Door and window schedule.

drawings. Common features include the following:

- A reference mark, letter, or number that corresponds to the markings on the drawings
- The desired manufacturer for a specific item
- Information on item part numbers, sizes, special finishes, and hardware requirements

Door and window schedules are designated by a number or letter on the drawings. The same letter or number is duplicated in the schedule, with a brief description of the item. Typical door and window schedules must show their relationship to the floor plan.

A finish schedule (*Figure 13*) references a location by a room number noted on the schedule; the drawing usually references the schedule with a symbol or note. Finish schedules for commercial buildings provide detailed lists of finishing materials and their application, plus installation procedures for floors, walls, base trim, ceilings, and molding.

Certain areas of a floor plan, elevation, or other drawing may be enlarged for greater clarity. These enlargements are drawn to a larger scale. They are called details (*Figure 14*). Details can be found either on the sheet where they are first referenced, or grouped together on a separate detail sheet included in the set of drawings. These drawings are important sources of information for the contractor and craftsperson.

ROOM FINISH SCHEDULE

ROOMS	FLOOR				CEILING				WALL				BASE				TRIM			REMARKS
	CARPET	CERAMIC TILE	RUBBER TILE	CONCRETE	ACOUSTIC TILE	DRYWALL	PAINT	CERAMIC TILE	DRYWALL	PAINT	WALLPAPER	CERAMIC TILE	WOOD	RUBBER	CERAMIC TILE	STAIN	WOOD	STAIN	PAINT	
ENTRY		✓			✓				✓	✓	✓		✓			✓	✓	✓		See owner for all painting
HALL	✓				✓				✓	✓			✓			✓	✓	✓		
BEDROOM 1	✓				✓				✓	✓	✓		✓			✓	✓	✓		See owner for grade of carpet
BEDROOM 2	✓				✓				✓	✓			✓			✓	✓	✓		See owner for grade of carpet
BEDROOM 3	✓				✓				✓	✓			✓			✓	✓	✓		See owner for grade of carpet
BATH 1	✓	✓			✓			✓	✓	✓	✓	✓	✓			✓	✓	✓		Wallpaper 3 walls around vanity
BATH 2		✓			✓			✓	✓	✓	✓	✓			✓		✓	✓		Water-seal tile Wallpaper w/wall
UTIL + CLOSETS	✓		✓			✓	✓		✓	✓			✓			✓	✓	✓	✓	Use off-white flat latex
KITCHEN			✓		✓				✓	✓				✓			✓	✓		
DINING	✓				✓				✓	✓	✓		✓			✓	✓	✓		
LIVING	✓				✓				✓	✓			✓			✓	✓	✓		See owner for grade of carpet
GARAGE				✓		✓	✓					✓		✓			✓	✓		

201F13.EPS

Figure 13 ◆ Finish schedule.

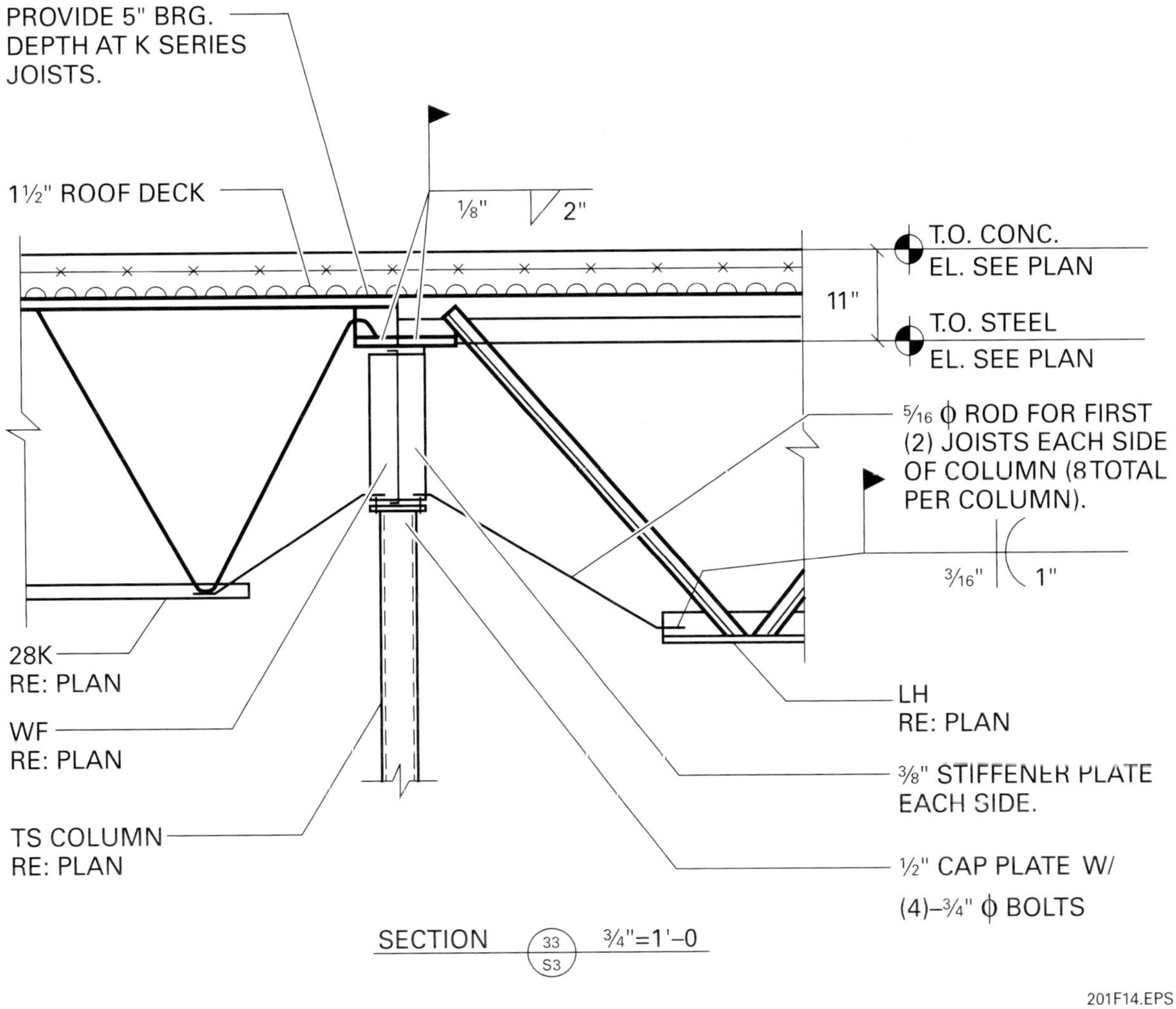

Figure 14 ◆ Detail drawing.

3.1.4 Elevations and Sections

Elevation views on commercial construction drawings are similar to residential elevations, but provide more information. Exterior elevations (*Figure 15*) provide views of the building from each major orientation, as well as references for section views. Elevations are normally drawn to the same scale as the floor plans.

Some interior elevations provide vertical dimensioning for interior work, materials lists, and construction details. They are important for built-in cabinets, shelving, finish carpentry, or millwork items. The scale of the drawing depends on the detail required, and may be as small as ¼ inch to 1 foot, or as large as ¾ inch to 1 foot.

There are many wall sections to show the different types of exterior and interior walls. The interior sections detail the construction of each wall type, such as curtain walls, partitions, loadbearing walls, fire-resistant walls, and noise-reduction walls. The floor plans incorporate a legend that specifies the interior wall type. The wall section drawing provides the necessary detail. Wall sections usually provide the following details:

- Construction techniques and materials types
- Stud types and placement
- Fire ratings of various materials measured in terms of hours of resistance
- Sound barrier placement or materials
- Insulation applications and materials
- In-wall features such as recesses or chases

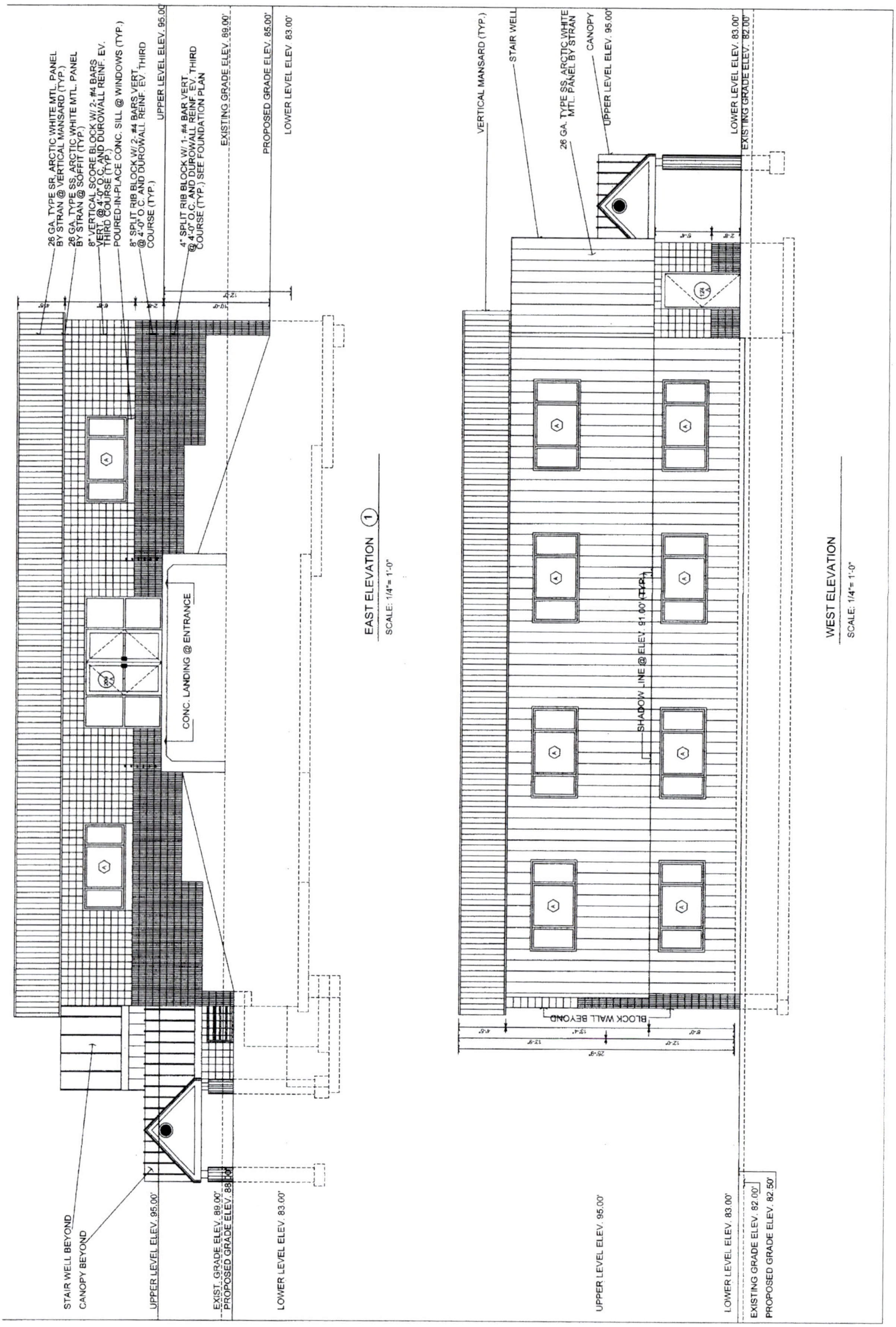

201F15.EPS

Figure 15 ◆ Elevation drawings.

Elevations and sections are also drawn for landings, stairways, and backfilled retaining walls. The section drawings detail construction techniques or materials. For instance, a stairway will have details, sections, and elevations with information on tread, landing, and handrail construction. The amount of detail will depend on the complexity of the stairway and on the various **building code** specifications.

Longitudinal and transverse wall sections show construction features that are expanded on the structural plans. These sections will show the following:

- Relationships of all wall features from the footings through the roof
- Footing and foundation placement in relation to other elevations
- Exterior materials symbols and notations
- Framework type and placement

As with the elevations, the sections provide an overall view of the proposed structure rather than the detail required for construction. Detail notations or callouts refer directly to the structural plan sheets. Most callouts refer to details on roofing **beams** and trusses, foundations, framing features, or other structural components.

3.2.0 Structural Drawings

Structural drawings (*Figure 16*) provide detailed information on the structural features of the building. This includes information on the loadbearing design and materials, such as masonry, reinforced concrete, or steel framing. For example, if the floor, roof beams, or trusses place their weight directly on the wall materials, the structure has loadbearing walls. This means that the exterior walls are constructed of materials with high **compressive strength** such as brick, block, or cast concrete. The walls support their own weight as well as that of the various floor and roof elements. The plans for such loadbearing walls will show no separate supporting structural beams or columns along the wall except for thickened wall sections that support major floor or roof beams. This type of construction is typical in smaller commercial buildings with spans of about 40 feet and heights of no more than three stories.

In reinforced concrete construction, the loadbearing elements usually include reinforced concrete footings, foundations, piers, beams, floors, columns, and pillars. Various sizes and quantities of carbon steel reinforcement bar (rebar) and/or steel tensioning cable are assembled inside the various building elements before concrete is placed to form the building element. Building exterior cladding is typically non-bearing curtain walls of masonry or other materials.

High-rise buildings are typically steel frame construction with non-bearing exterior curtain walls. The framework for the entire structure is formed by bolting or welding various steel elements together. The loads from floors and roof are transferred to beams and **girders** and down steel columns to the footings. When the joints are bolted, the project schedules designate the number, size, and material requirements for the bolts. *Figure 17* shows the most common shapes for steel frame elements and lists the typical plan designations.

The structural drawings include plan views, sections, details, schedules, and notes. They provide information on the size and placement of loadbearing elements. They also show how the loadbearing elements are connected to each other and to other parts of the structure.

Typical structural drawings include a foundation plan, floor framing plans, and a roof framing plan (*Figure 18*). The plan view will have sections, details, schedules, and notes located in any available space on the drawing sheet. Each plan view should have a north directional arrow to maintain a consistent orientation. Plan views are typically drawn to the scale of $\frac{1}{8}$ or $\frac{1}{4}$ inch to 1 foot; sections are $\frac{1}{2}$ or $\frac{3}{4}$ inch to 1 foot; details are 1 or $1\frac{1}{2}$ inch to 1 foot.

INSIDE TRACK

Check the Legend

To avoid mistakes in reading the drawings, be sure you understand the symbols and abbreviations used on every drawing set. Symbols and abbreviations may vary widely from one drawing set to another.

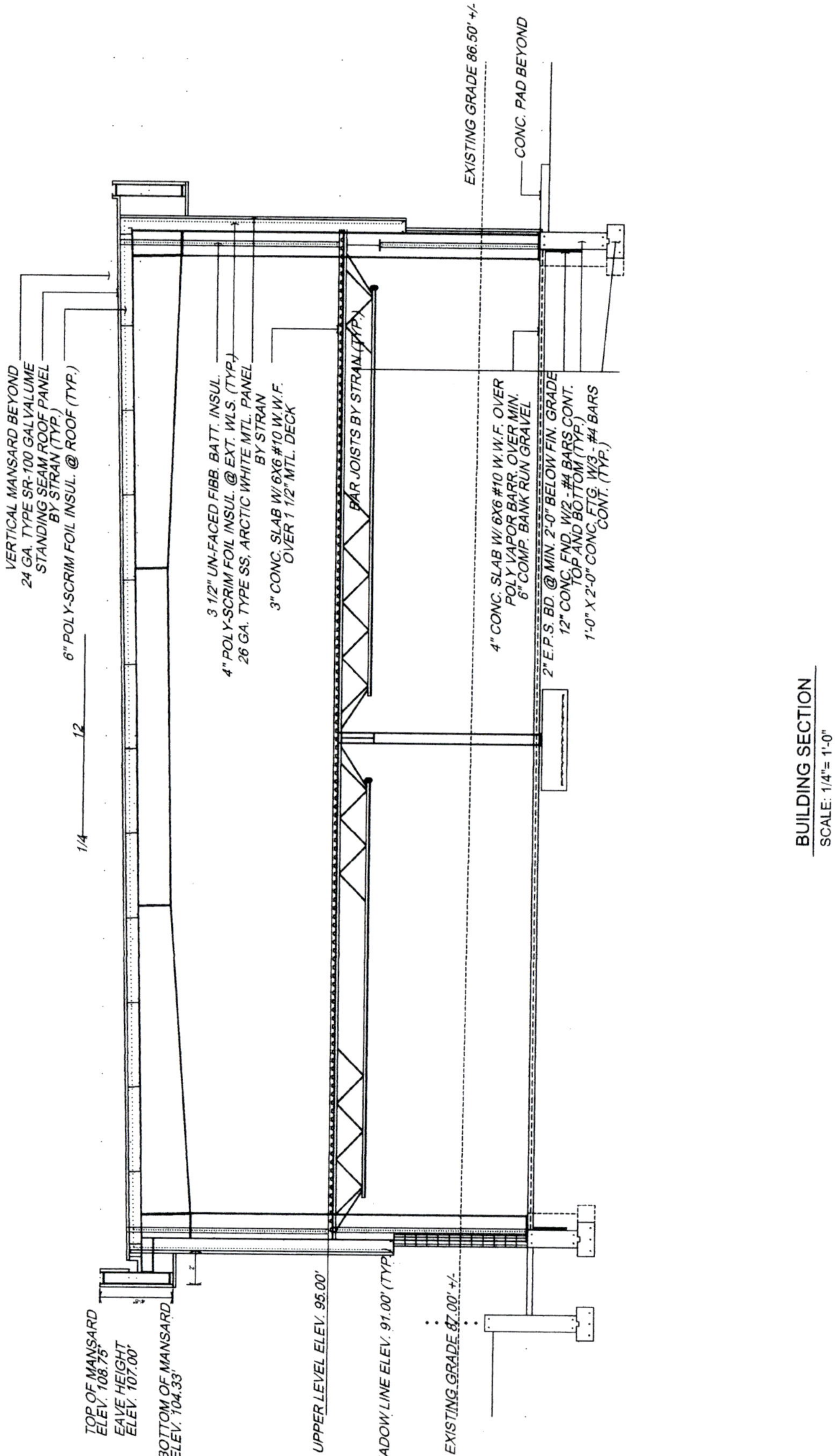

201F16.EPS

Figure 16 ◆ Structural drawing.

Descriptive Name	Shape	Identifying Symbol	Typical Designation height/wt/ft in lb	Nominal Size height width
Wide Flange Shapes	I	W	W21 × 132	21 × 13
Miscellaneous Shapes	I	M	M8 × 65	8 × 2¼
American Standard Beams	I	S	S8 × 23	8 × 4
American Standard Channels	[	C	C6 × 13	6 × 2
Miscellaneous Channels	[	MC	MC8 × 20	8 × 3
Angles Equal Legs	L	L	L 6 × 6 × ½	6 × 6
Angles Unequal Legs	L	L	L 8 × 6 × ½	8 × 6
Structural Tees (cut from wide flange)	T	WT	WT12 × 73	9
Structural Tees (cut from Am. std. beams)	T	ST	ST9 × 35	9

201F17.EPS

Figure 17 ◆ Structural steel notations.

Each plan is referenced to a grid or checkerboard identifying the placement of columns and/or footings. The grid is the same on all plans. As noted in the section on floor plans, the grid also shows dimensioning lines. In the structural drawings, a callout sequence number or mark will identify and show the placement of columns on the grid. Some project plans do not use a grid, but use a callout sequence marking. In either case, the identification marks will be referenced to schedules or notes.

3.2.1 Structural Framing Plans

The structural engineer draws a framing plan or diagram for the roof and for each floor level that will be framed. On these drawings, the exterior walls or bearing walls are often lightly drawn in while heavier lines represent the framing. The resulting plan looks very much like a graph or diagram, as shown in *Figure 19*.

Notice the following when looking at this diagram, or at any framing plan:

- Notations identifying beams, joists, and girders by size, shape, and material
- Column, pier, and support locations and their relationship to joists or framing
- Notes or callouts identifying corresponding sections or detail drawings

You may also see details for locations of stairs, recesses, and chimneys. Details show additional unique framing around these areas. The dimensions on the drawings are center line dimensions, not actual member dimensions. The member size must be less than the center line dimensions to allow for construction tolerances.

The columns on a framing plan are shown from the top. Lines running between columns are beams. Beams fasten directly to the columns. Joists fasten between beams or between beams and walls. To keep long spans from swaying or twisting in the center, bridging or support members are placed between joists. All of these members will either be referenced to a schedule or to notes directly written on the plans.

Structural drawings include sheets of details, schedules, and notes with the framing plans.

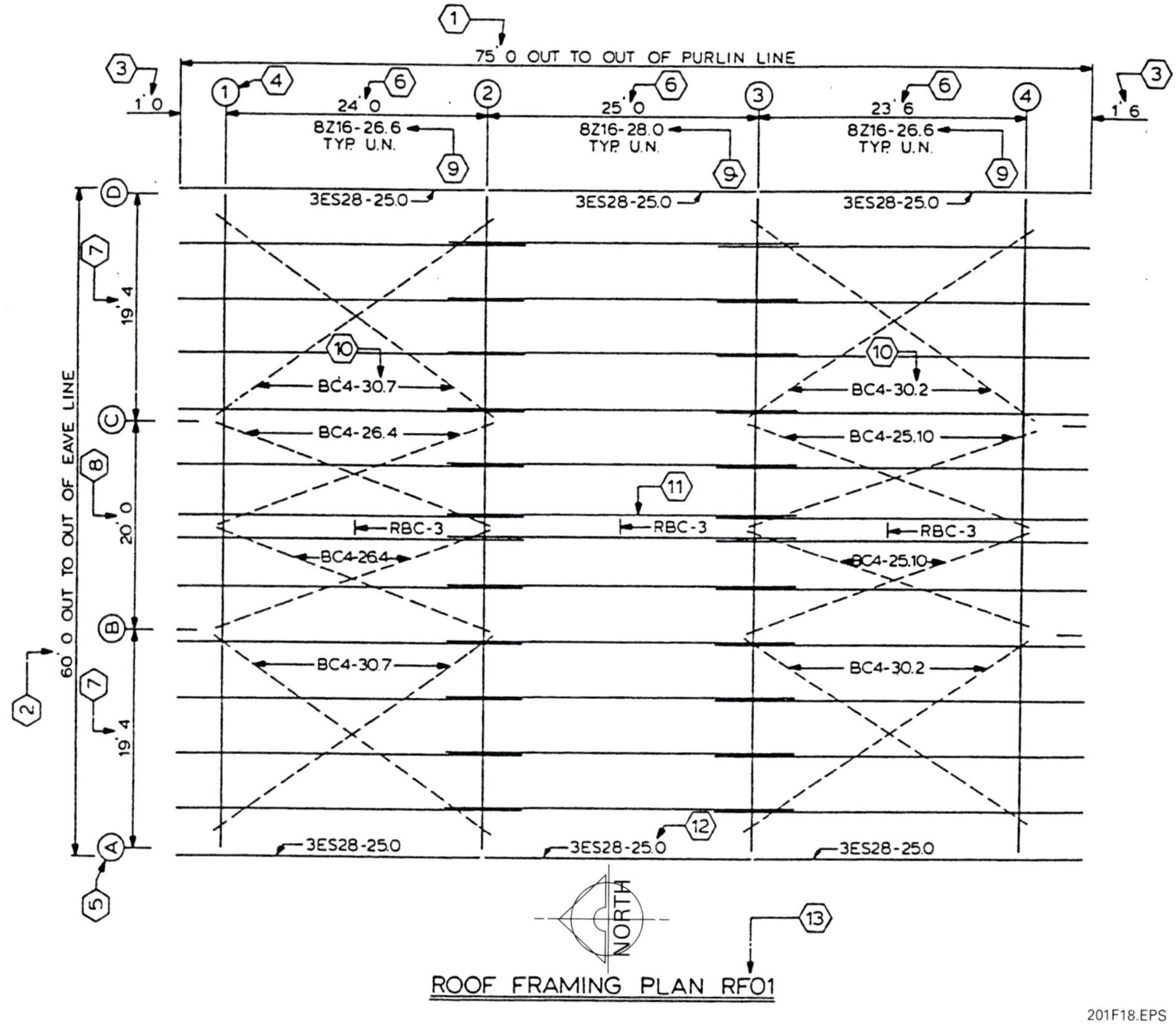

Figure 18 ◆ Roof framing plan.

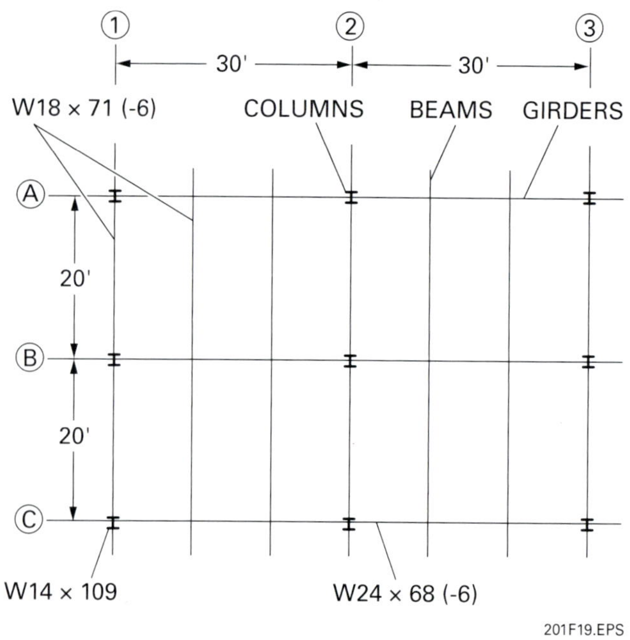

Figure 19 ◆ Structural steel framing diagram.

They help workers to understand and follow the specifications for the structure, and provide information on the following areas:

- Reinforcing information for all areas where steel rods or wire mesh will be used
- Information for each type of connection made in framing members
- Bearing plate information detailing the features of all members which will bear directly on other members
- Information for positioning ties, stirrups, or saddles
- Placing and constructing information for any unique features that cannot be adequately described with a drawing

The details also identify load limits, test strength, fastener types, and uniform specifications, which must be applied where specific information is not given.

There is a great deal of information on the framing plans. It is easier to read such plans by isolating the separate bays or spans between columns. Read the details for that area before moving to other areas of the plan.

3.2.2 Concrete Structures

Concrete is one of the most widely used materials in construction today. Concrete's strength and durability make it a desirable construction material. Concrete is typically used in building foundations because it has a high compressive strength. Concrete on its own has low **tensile strength,** which means that concrete can be easily broken apart when it is flexed. When reinforced with steel reinforcing bar, often called rebar, concrete has good tensile strength.

Concrete foundations typically have footings around the edges and wherever support columns are needed. Footings support the weight of the building walls and roof, so they need to be stronger than the building floor. Footings have more steel reinforcement than foundations and are dug much deeper into the ground, usually below the regional frost line. *Figure 20* shows partial plans for a building with concrete walls, footings, and floor.

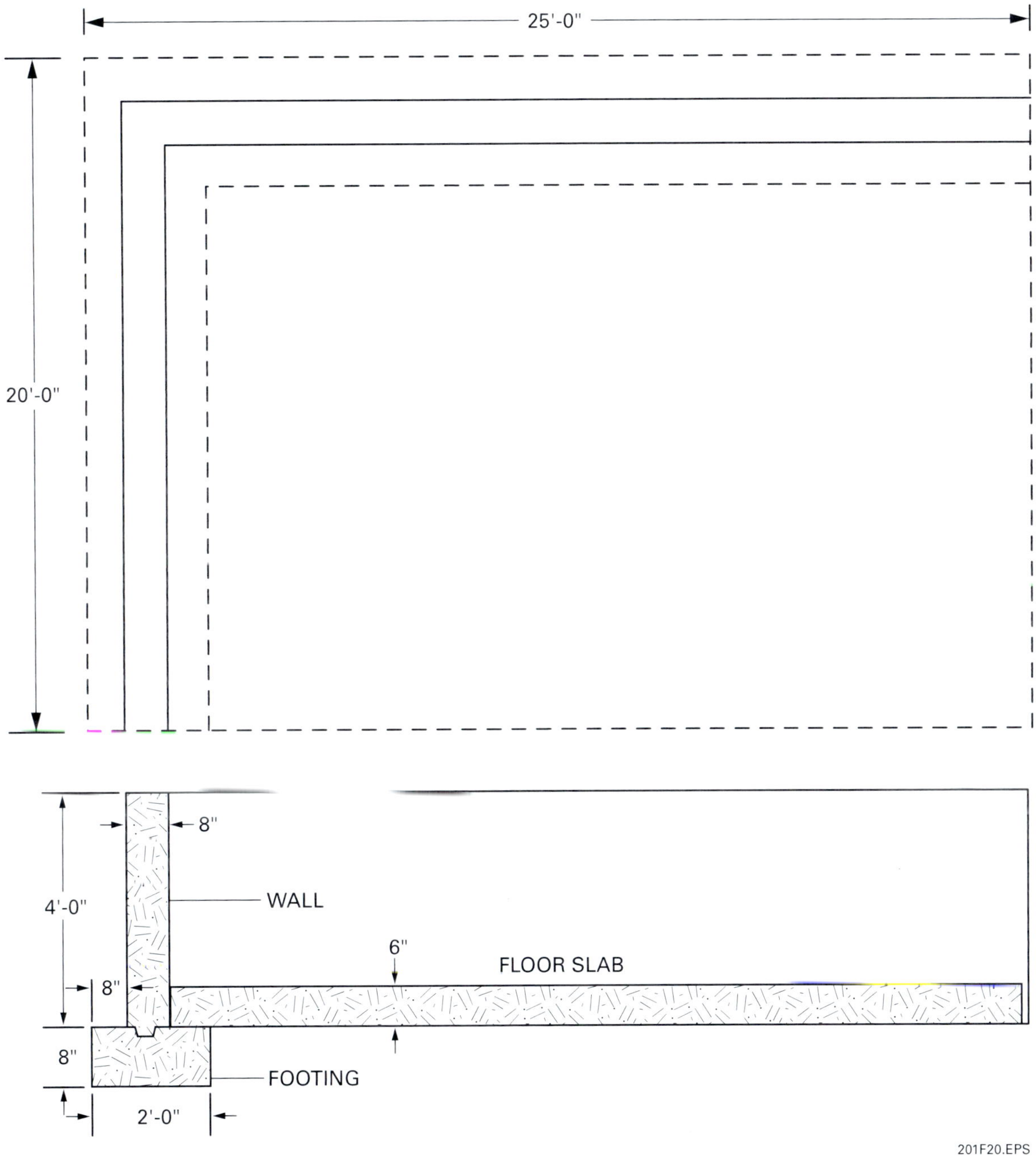

Figure 20 ◆ Partial wall, footing, and floor slab plan.

Frequently, the framework for a large construction project will need to use special, heavy-duty reinforcing techniques to strengthen the concrete. When a load is applied to a conventional concrete slab, such as an elevated parking garage deck, the concrete tends to sag and may develop cracks. The use of rebar in the slab combats this problem, but is not enough to prevent cracking under heavy loads. To help correct this, post-tensioned and pre-tensioned reinforcement is used.

In post-tensioning, a steel tendon is placed in the form with the ends protruding. The tendon is placed in accordance with a post-tensioning profile drawing developed by a qualified engineer. *Figure 21* is an example of post-tensioning profiles for post-tensioned beams. The multi-span beam profile diagram shows how the tendons go across the top of a column. A tendon consists of a threaded bar or strands of wire, along with anchoring hardware and sheathing. The strand is typically made from seven steel wires, each ½ inch in diameter, that have been twisted together. Once the concrete has hardened around the tendon, one end is anchored, and the other end is tensioned using a special hydraulic jack. When the desired tension has been reached, the tendon is secured.

The two types of post-tensioning are bonded and unbonded. In bonded post-tensioning, a steel or plastic duct is inserted in the form. After the concrete has been poured and has hardened, the strand is threaded through the duct. Tensioning is applied, and then the duct is filled with grout.

Figure 22 shows post-tensioning tendons protruding from recently poured concrete. Once the tendons have been anchored, they are cut, and the openings are sealed (*Figure 23*).

In an unbonded system, the strand is covered with corrosion-inhibiting grease and encased in a

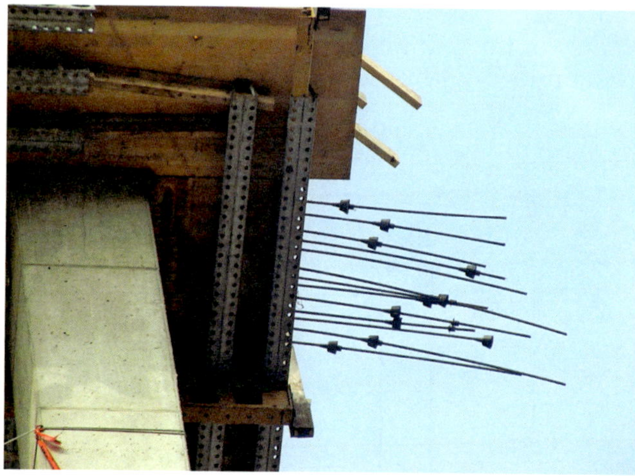

201F22.EPS

Figure 22 ◆ Post-tensioning tendons.

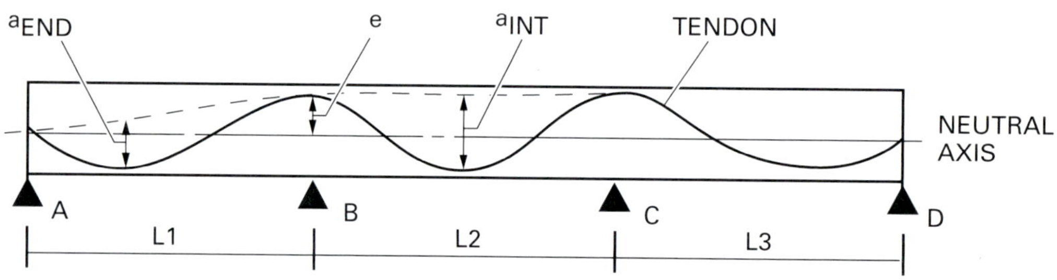

SINGLE-SPAN BEAM

201F21A.EPS

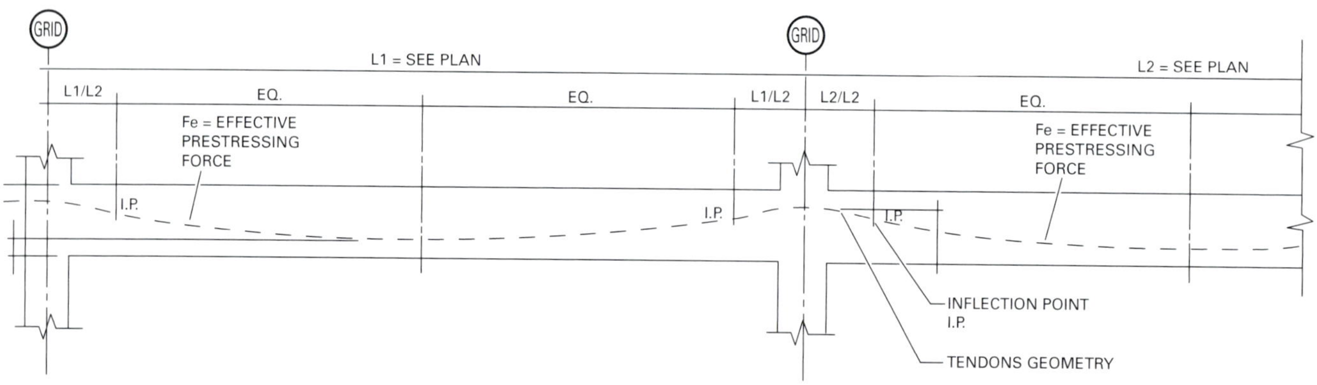

MULTI-SPAN BEAM

201F21B.EPS

Figure 21 ◆ Examples of tendon profiles.

201F23.EPS

Figure 23 ◆ Sealed tendon openings.

waterproof plastic sheath. The entire assembly is placed into the form before the concrete is poured.

The twisted-wire strand is used in large structures. The threaded bar is more common in smaller structures; a bearing plate and nut are used to anchor the bar.

WARNING!

Do not drill or install powder-actuated fasteners into a post-tensioned beam because you might damage the tendon, which may fly out of the slab, causing injury. Damaged tendons also reduce the loadbearing capacity of the beam or slab.

Pre-tensioning is the same as post-tensioning with this exception: with pre-tensioning, the elements of the entire structure are manufactured and tensioned at the manufacturing site, and then it is transported to the construction site for assembly.

The interior walls of concrete construction are typically framed with steel studs, but furring strips or full-size wood studs may also be used. These allow a means to install thermal and moisture barriers, electrical wiring, and drywall panels.

Cold-formed structural steel balloon framing is sometimes used on reinforced concrete or structural steel buildings to support certain types of curtain walls. In these cases, shop drawings for the balloon framing are prepared to detail the framing and its method of attachment to the building. At a minimum, the framing must be engineered to withstand wind loading and the weight of the exterior cladding system. *Figure 24* shows examples of such shop drawings for the use of balloon framing on a structural steel building.

3.3.0 Mechanical Drawings

Mechanical drawings provide information about the HVAC systems, plumbing systems, and fire protection systems. The work required to install these systems in commercial structures is typically performed by trades or firms that specialize in the particular craft.

While drywall technicians generally do not get involved in with working on these systems, there are many situations in which you will be required to make passage for or work around mechanical items. One example would be the through-wall ductwork for a forced-air HVAC system. Because of these types of requirements, you must carefully review all the mechanical drawings for information on dimensions and measurements. Locate services entering the building and passing through interior walls and coordinate with the HVAC workers.

INSIDE TRACK

Coordination Drawings

Coordination drawings are produced by the individual contractors for each trade to prevent a conflict in the installation of their materials and equipment. Coordination drawings are produced prior to finalizing shop drawings, cut lists, and other drawings, and before the installation begins. Development of these drawings evolves through a series of review and coordination meetings held by the various contractors.

Some contracts require coordination drawings, while others only recommend them. When one contractor elects to make coordination drawings and another does not, the contractor who made the drawings may be given the installation right-of-way by the presiding authority. As a result, the other contractor may have to bear the expense of removing and reinstalling equipment if it was installed in a space designated for use by the contractor who produced the coordination drawings.

1.00 DESIGN CRITERIA

BUILDING CODE: 2003 IBC
BASIC WIND SPEED: 110 MPH (3 SEC GUST)
EXPOSURE: "B" IF: 1.15
MEAN ROOF HEIGHT: 66'-0"

NO OTHER LOADS CONSIDERED

DEFLECTION LIMITS: L/360

2.00 EXTERIOR WIND LOAD

P := -28.0 PSF INTERIOR ZONES

P := -32.5 PSF CORNER ZONES

CORNER ZONE IS A DISTANCE OF 10'-0" FROM WALL CORNERS

3.00 PRODUCT IDENTIFICATION

PRODUCT IDENTIFICATION

600S162-43

- 600 — MEMBER DEPTH
- S — MEMBER STYLE
- 162 — FLANGE WIDTH
- 43 — MEMBER THICKNESS (MIL)

MATERIAL THICKNESS COMPARISON TABLE

MILS	GAUGE	MINIMUM DELIVERED THICKNESS	DESIGN THICKNESS
33	20	0.0329"	0.0346"
43	18	0.0428"	0.0451"
54	16	0.0538"	0.0566"
68	14	0.0677"	0.0713"
97	12	0.0966"	0.1017"

SECTION	SSMA IDENTIFICATION	FLANGE WIDTH
S-SECTIONS (STUDS WITH ½" STIFFING LIPS)	S137	1⅜"
	S162	1⅝"
	S200	2"
	S250	2½"
T-SECTIONS (TRACKS)	T125	1¼"
	T200	2"
U-SECTIONS (U-CHANNEL)	075U50	¾"
	150U50	1½"
	200U50	2"
	250U50	2½"
F-SECTIONS (F-CHANNEL)	087F125	⅞"
	150F125	1½"

4.00 MATERIAL NOTES

4.01. ALL STUDS AND JOIST MEMBERS 16 GAUGE AND HEAVIER SHALL BE FORMED FROM STEEL CORRESPONDING TO A TYPE LISTED IN THE A.I.S.I. "SPECIFICATION FOR THE DESIGN OF COLD-FORMED STEEL STRUCTURAL MEMBERS", WITH A MINIMUM YIELD STRENGTH OF 50 KSI UNLESS SPECIFICALLY NOTED OTHERWISE.

4.02. ALL 18 GAUGE AND LIGHTER MEMBERS, AND ALL TRACK, BRIDGING, AND ACCESSORY ITEMS SHALL BE FORMED FROM STEEL "SPECIFICATION FOR THE DESIGN OF COLD-FORMED STEEL STRUCTURAL MEMBERS", WITH A MINIMUM YIELD STRENGTH OF 33 KSI UNLESS SPECIFICALLY NOTED OTHERWISE.

4.03. STRUCTURAL PROPERTIES AND CAPACITIES OF STEEL FRAMING COMPONENTS SHALL BE IN ACCORDANCE WITH THE A.I.S.I. "SPECIFICATION FOR THE DESIGN OF COLD-FORMED STEEL STRUCTURAL MEMBERS".

LAVI & ASSOCIATES, L.P. CONSULTING ENGINEERS 10300 N. CENTRAL EXPRESSWAY, SUITE #322 DALLAS, TX 75231 T: (214)340-0049 F: (214)340-0067	DECOUPLING OF THE 1954 BUILDING ST. LUKE'S EPISCOPAL HOSPITAL HOUSTON, TEXAS FOR: MAREK BROTHERS SYSTEMS, INC. HOUSTON, TEXAS	JOB 08-373 DATE 10/27/08	SHEET LSF1.0

201F24A.EPS

Figure 24 ◆ Shop drawing for balloon framing (1 of 4).

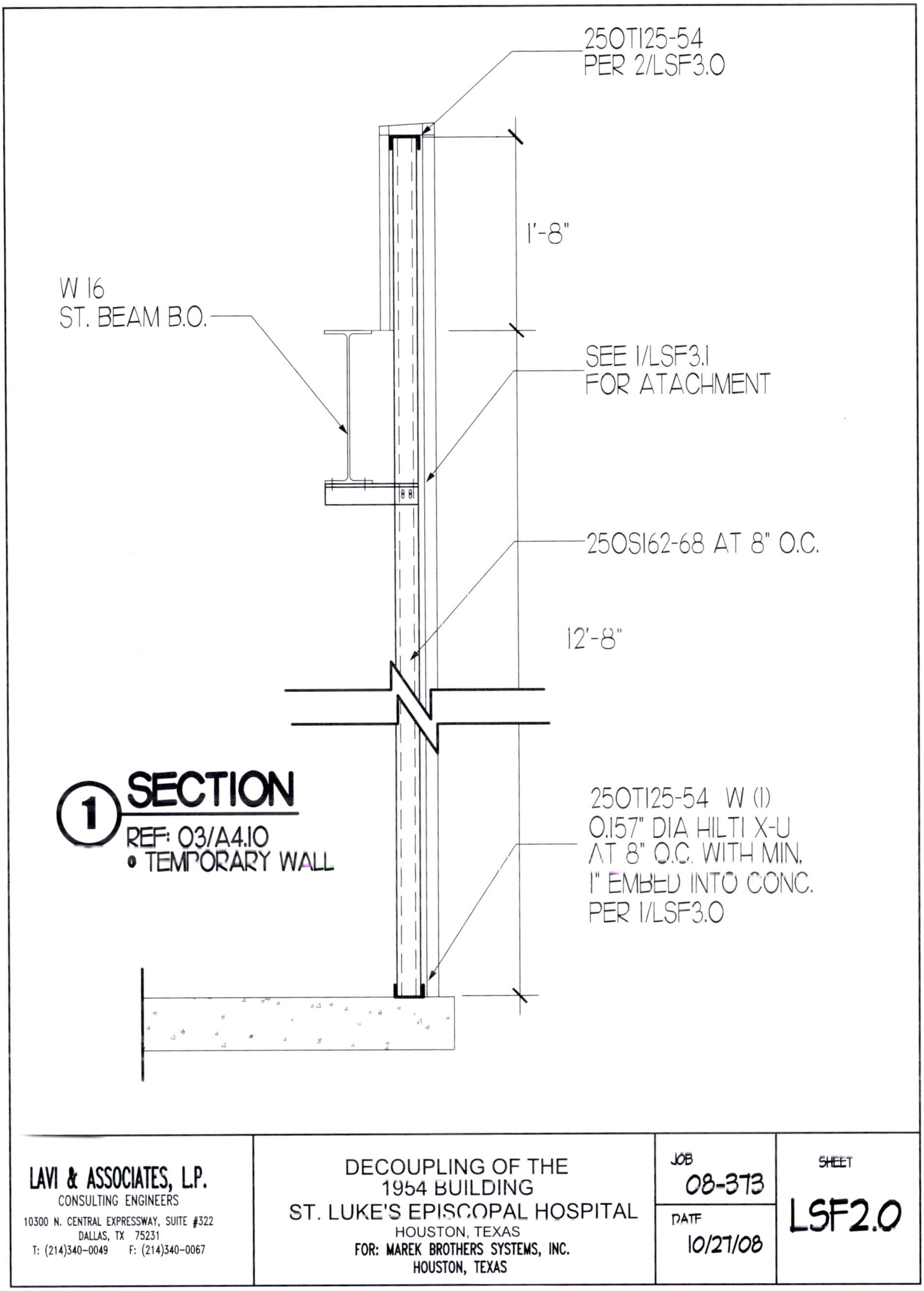

201F24B.EPS

Figure 24 ◆ Shop drawing for balloon framing (2 of 4).

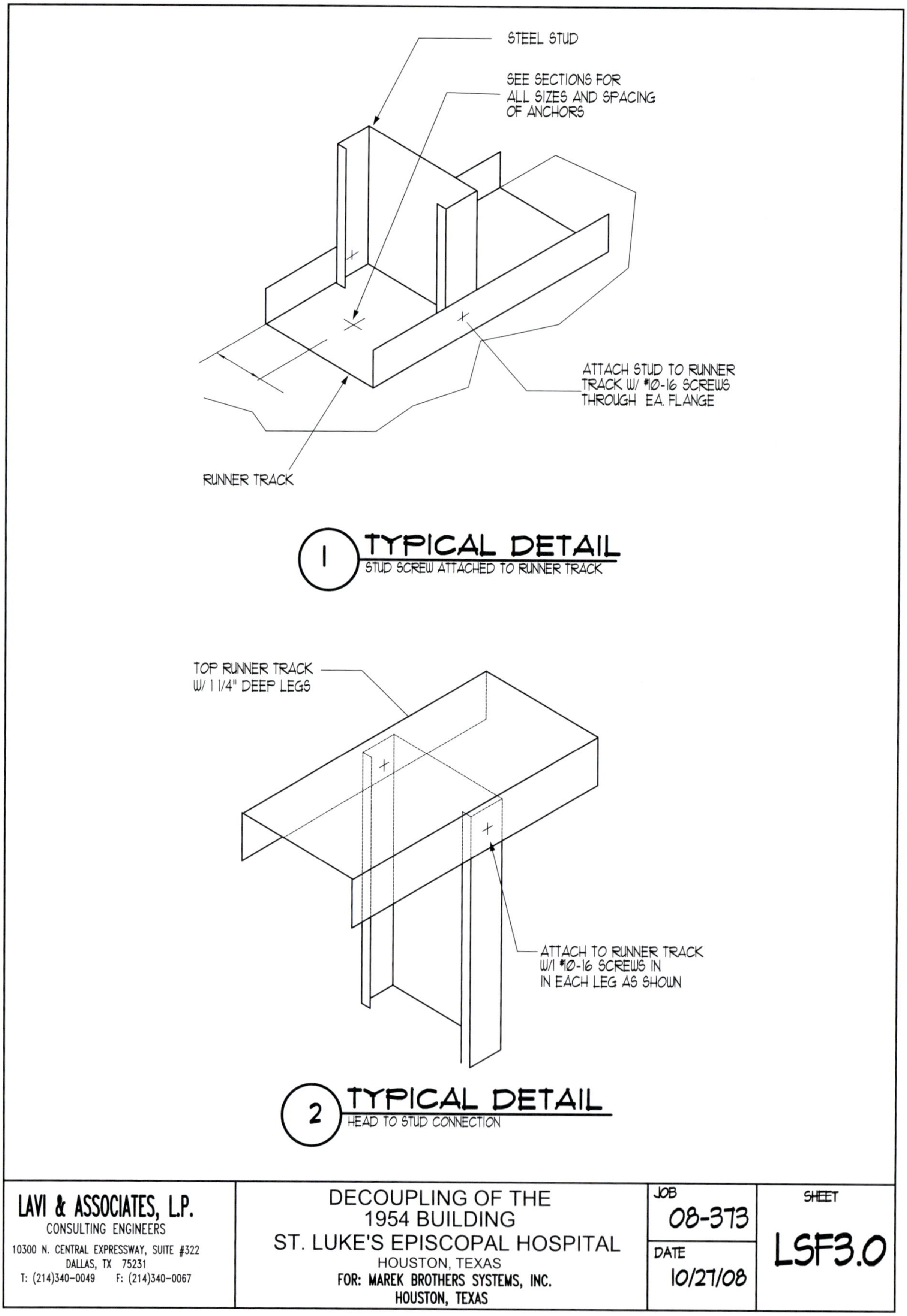

201F24C.EPS

Figure 24 ◆ Shop drawing for balloon framing (3 of 4).

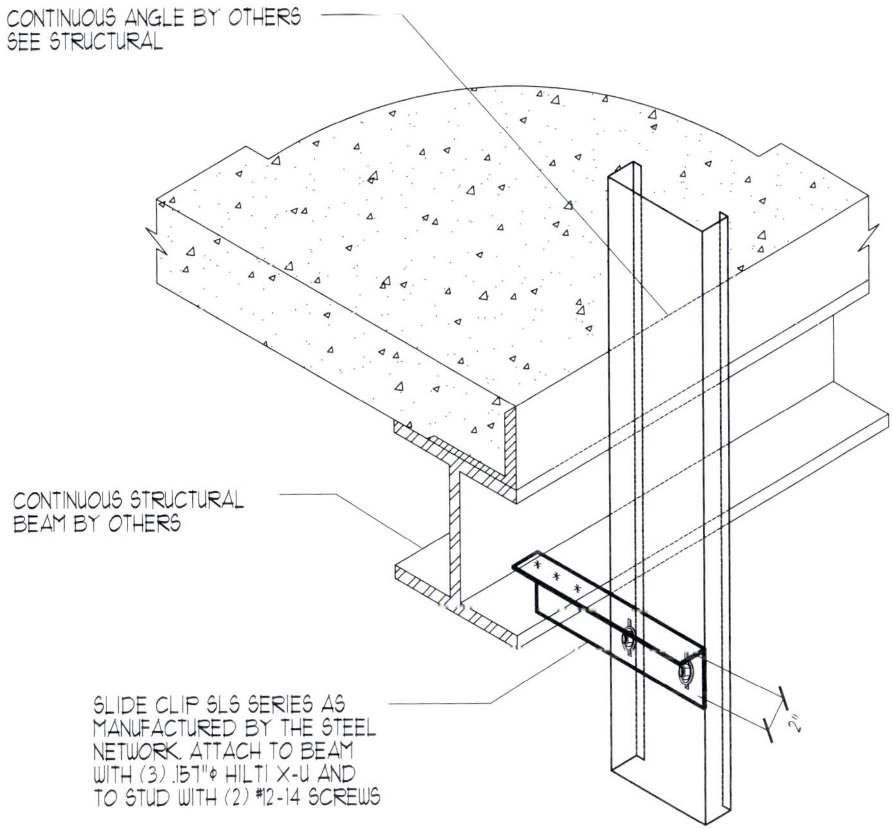

(1) TYPICAL DETAIL AT STUDS

CURTAIN WALL DEFLECTION
OFFSET BY-PASS CONDITION

LAVI & ASSOCIATES, L.P. CONSULTING ENGINEERS 10300 N. CENTRAL EXPRESSWAY, SUITE #322 DALLAS, TX 75231 T: (214)340-0049 F: (214)340-0067	DECOUPLING OF THE 1954 BUILDING ST. LUKE'S EPISCOPAL HOSPITAL HOUSTON, TEXAS **FOR: MAREK BROTHERS SYSTEMS, INC.** **HOUSTON, TEXAS**	JOB 08-373 DATE 10/27/08	SHEET LSF3.1

201F24D.EPS

Figure 24 ◆ Shop drawing for balloon framing (4 of 4).

3.3.1 Plumbing Systems

All buildings that will be occupied require some type of plumbing. Warehouses have simple toilets, while hospitals and restaurants have sophisticated plumbing systems. As with fire protection, plumbing work requires special knowledge and training. Plumbing work is usually performed by a licensed craftsperson. A separate permit and inspection process are required.

Plumbing drawings (*Figure 25*) usually appear as plan view drawings and as **isometric drawings** called riser diagrams. The plan views show the horizontal distances or piping runs. The riser diagram shows vertical pipes in the walls. Plumbing drawings for water systems usually show two separate systems: a water source or distribution system and a waste collection and disposal system.

The incoming water systems operate under pressure. The distribution system shows the hot- and cold-water supply lines. Distribution plans usually include several of the following types of information:

- Piping type and size: types include copper, plastic, brass, iron, or steel piping; the sizes usually appear as nominal dimensions approximating the inside diameter (ID) of the pipe
- Pipe fitting at joints, which includes couplings (straight-run joints), elbows (45- or 90-degree bends), tees, and valves (also called cocks, bibbs, or faucets)
- Cold water supply piping from the street main to the service meter and on to the various fixtures
- Hot water supply piping locations from the hot water heater to the various fixtures, in many instances, includes a hot-water return line.
- A legend identifying cold- and hot-water supply lines

The waste system works using gravity, and it must be vented. The waste system is known as drain, waste, and vent, or DWV piping. When reading waste disposal plans, you can expect to see a combination of several notations. You will usually find the following information on a plumbing drawing for a waste disposal system:

- Waste disposal line size
- Location of fixture branches (horizontal disposal pipes)
- Stack locations (vertical disposal pipes are called waste stacks if they carry toilet waste; soil stacks if they carry waste materials other than toilet waste; and vent stacks if they extend through the roof to release gases within the stacks)

DID YOU KNOW?

The Empire State Building

The Empire State Building has 102 stories. It has 70 miles of water pipe that provide water to tanks at various levels. The highest tank is on the 101st floor. There are two public restrooms on each floor and a number of private bathrooms. There are, however, no water fountains.

- Locations for drains, sewer lines, cleanouts, traps, meters, and valves

Because waste disposal occurs using the force of gravity, fixture branches are sloped between ⅓ and ½ inch per foot away from the fixtures.

Piping for distribution of natural gas within the structure is often considered part of the plumbing work. The utility company installs a meter where natural gas enters the building. The piping starts at the meter. Natural gas is distributed to the gas-fueled appliances within the building such as boilers, furnaces, water heaters, and rooftop HVAC units.

Typically, black steel pipe with threaded fittings is used to distribute natural gas. The fittings are similar to those of other piping systems. They include elbows, bends, unions, and tees. Valves to control the flow of gas within the pipe are typically brass. Gas systems may require openings through masonry walls to accommodate piping, valves, and regulators. Through-wall piping must not be rigidly connected to the wall due to the different rates of expansion and contraction between the piping and the masonry.

3.3.2 HVAC Systems

Mechanical drawings that contain HVAC systems should be carefully studied for the layout, locations, and sizes of ductwork. Special mechanical drawings that illustrate the components of boiler and rooftop- or grade-mounted HVAC units may also be included.

Figure 26 includes three drawings for HVAC systems. The first shows a typical mechanical drawing for an HVAC system, which involves laying out all the piping, controllers, and air handlers to scale on a basic floor plan. Most of the details normally on a floor plan have been removed so the details of the HVAC system can be easily seen.

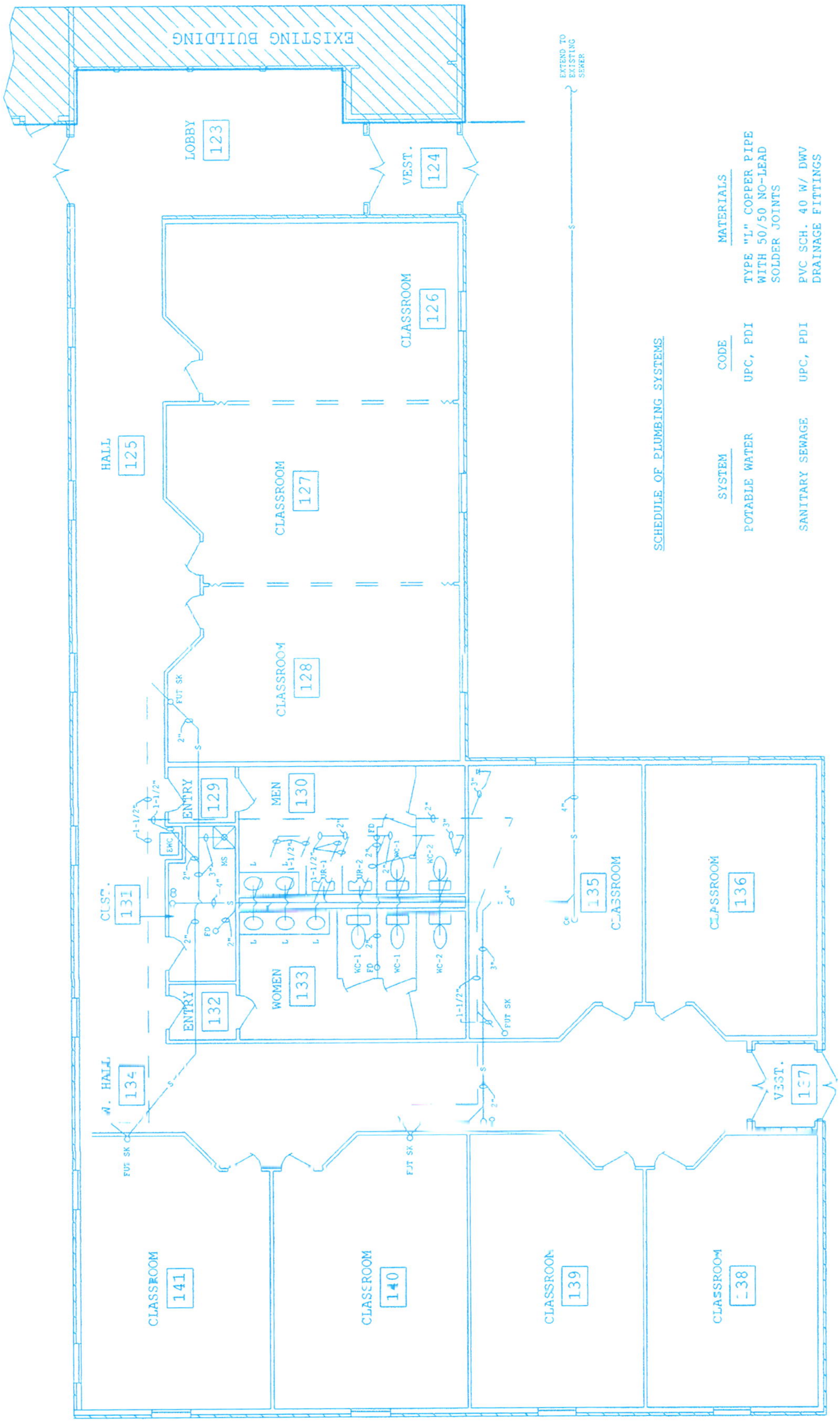

201F25A.EPS

Figure 25 ♦ Sanitary plumbing plan (1 of 2).

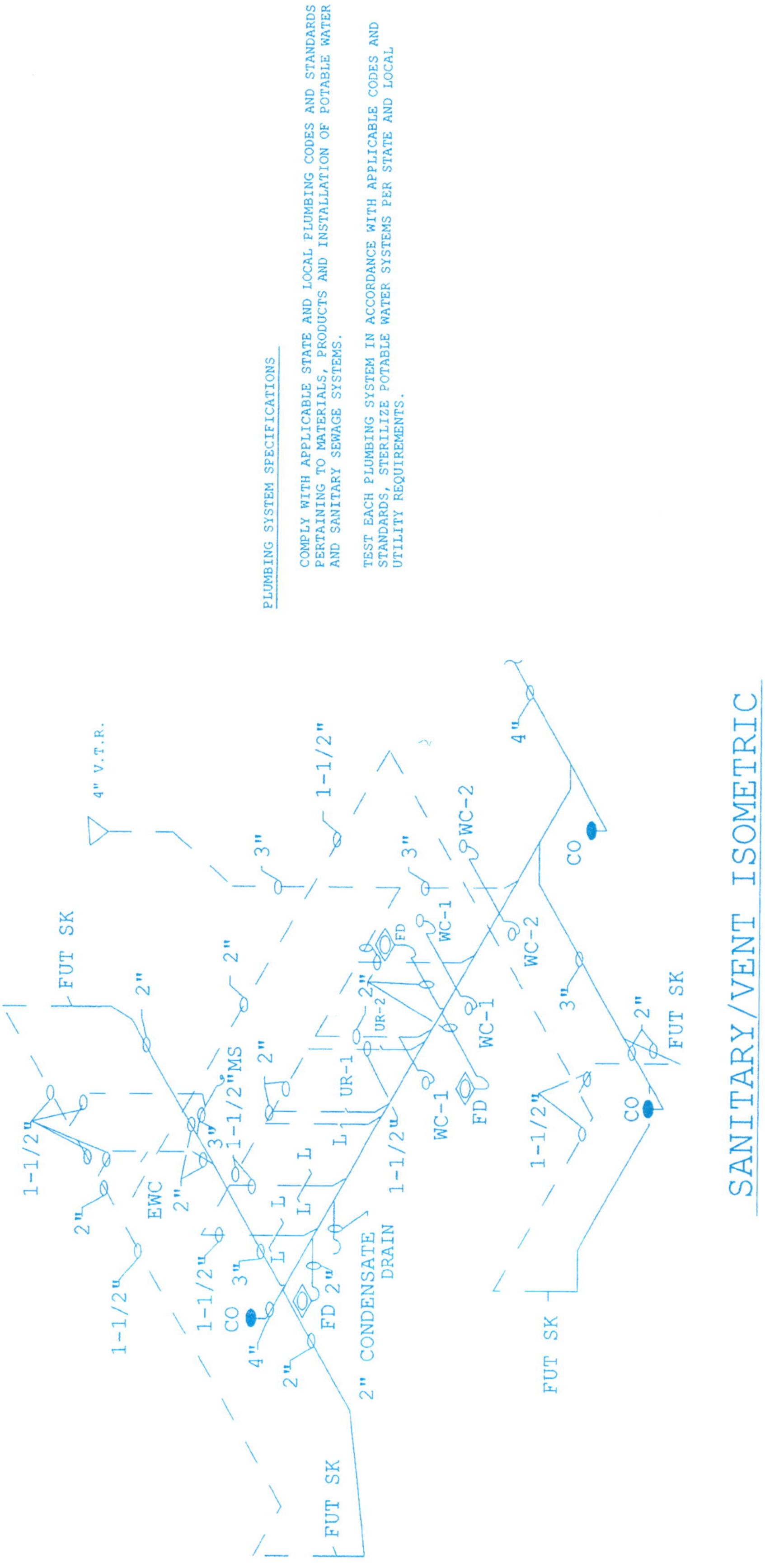

201F25B.EPS

Figure 25 ◆ Sanitary plumbing plan (2 of 2).

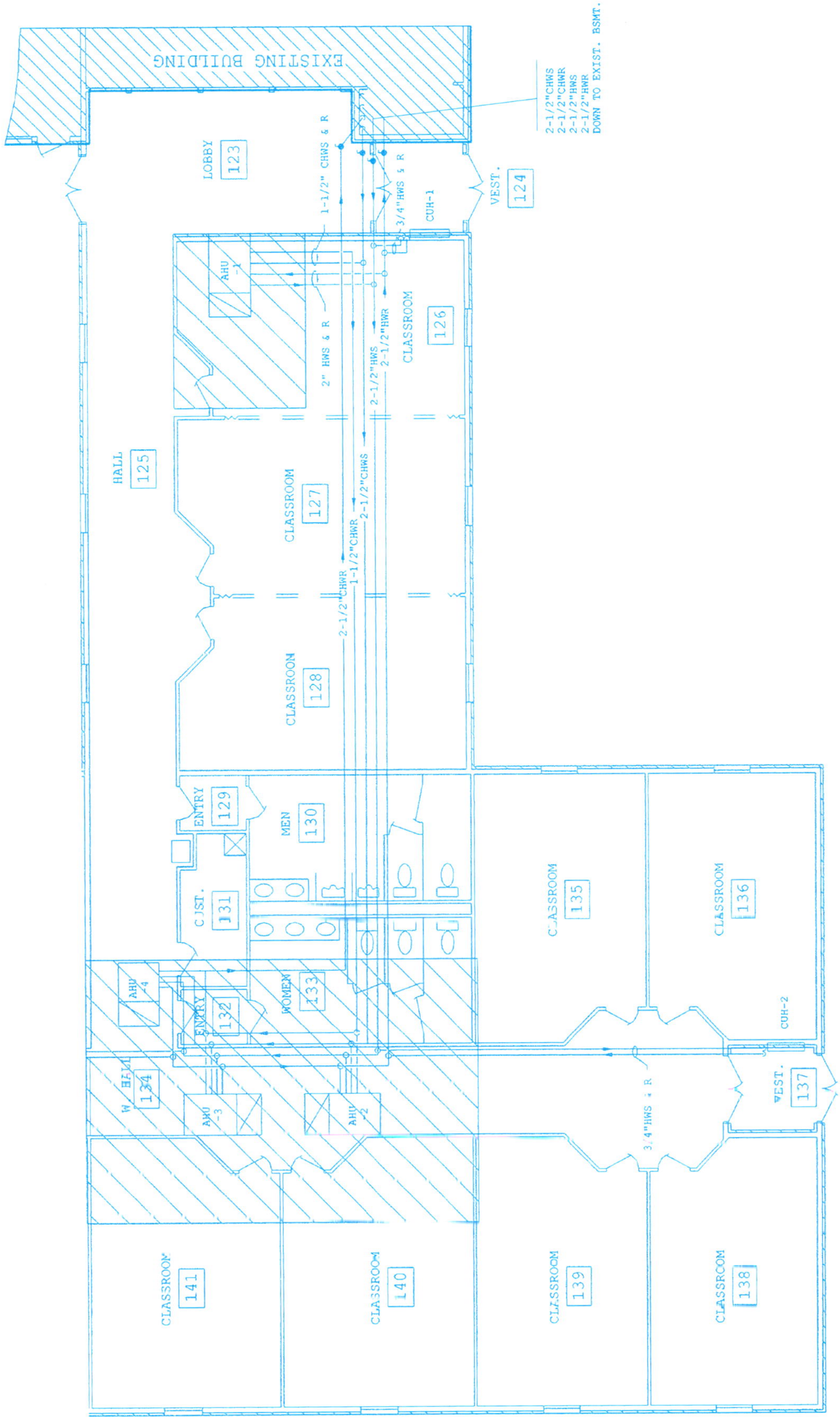

201F26A.EPS

Figure 26 ♦ HVAC mechanical plan (1 of 3).

INSIDE TRACK

Isometric Drawings

Isometric means *equal measurement*. A designer uses the true dimension of an object to construct the drawing. An isometric drawing shows a three-dimensional view of where the pipes should be installed. The piping may be drawn to scale or to dimension, or both. In an isometric drawing, vertical pipes are drawn vertically on the sketch, and horizontal pipes are drawn at an angle to the vertical lines.

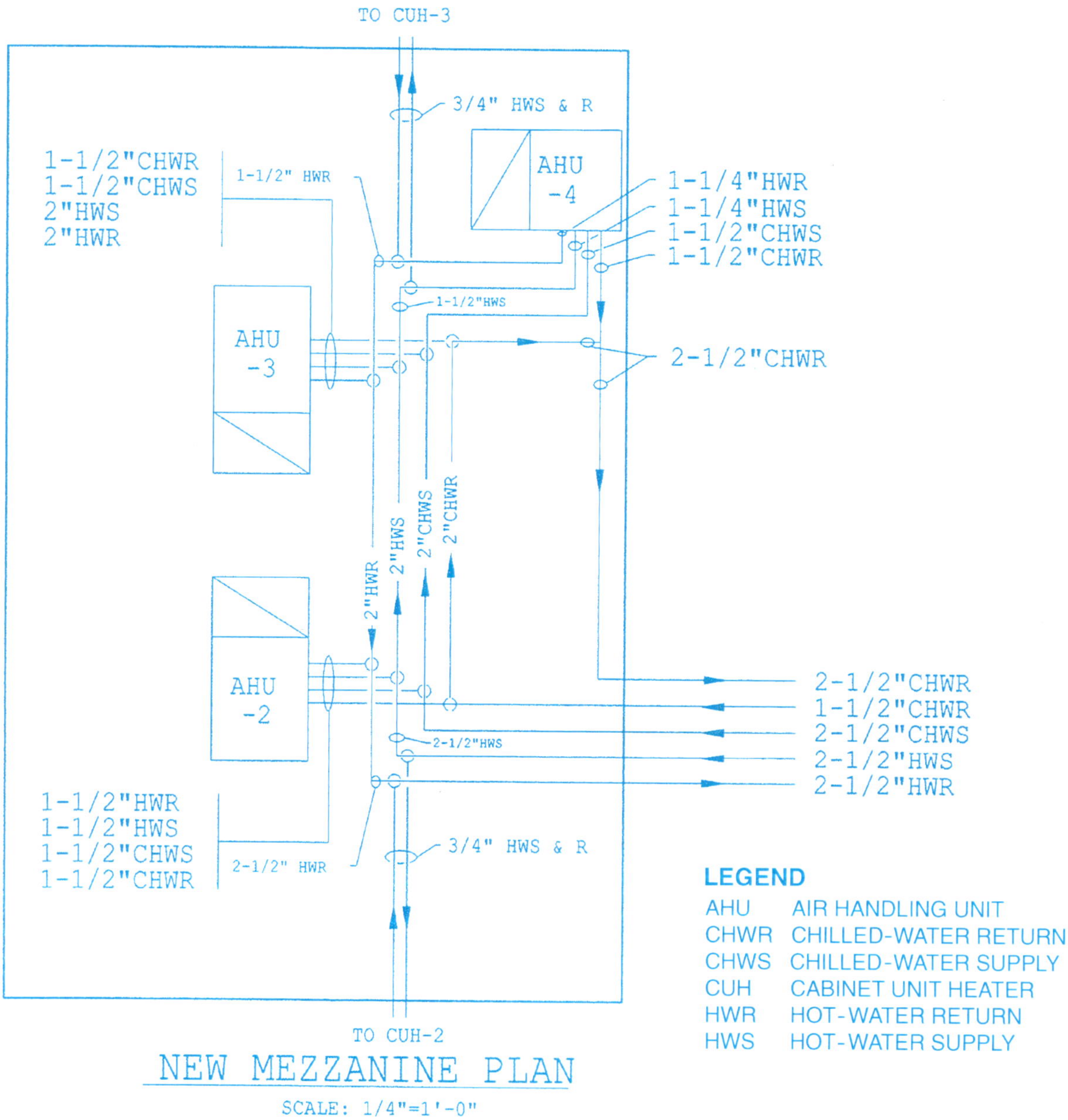

201F26B.EPS

Figure 26 ◆ HVAC mechanical plan (2 of 3).

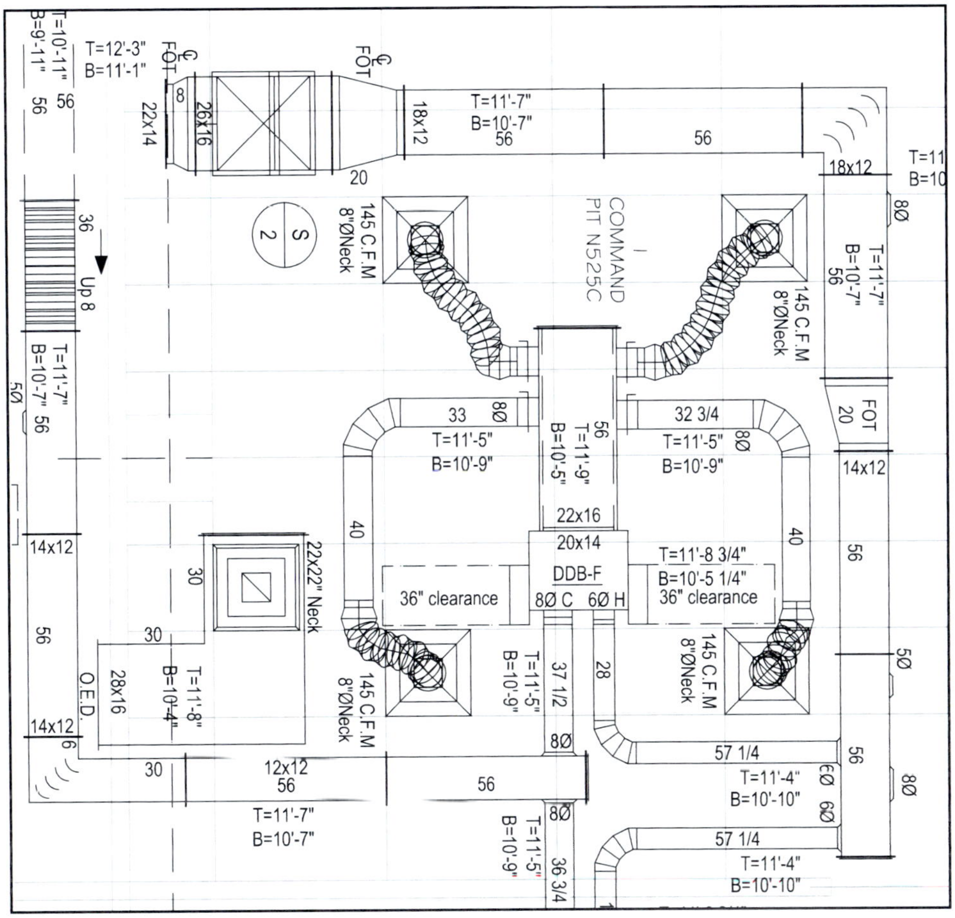
201F26C.EPS

Figure 26 ◆ HVAC mechanical plan (3 of 3).

The second drawing shown in *Figure 26* includes the following:

- Layout of all of the water supply and return and air handling units
- Piping diagrams for the hot-water and chilled-water coils
- Legend that identifies the abbreviations used in the drawings

Mechanical plans typically contain schedules that identify the different types of HVAC equipment. As appropriate, the plans will include a detailed view describing the installation of the HVAC equipment. Depending on the nature of the project, these views can include a refrigeration piping schematic, chilled-water coil and hot-water coil piping schematics, and piping runs for other HVAC equipment.

A basic understanding of the air handling system is a primary requirement for reading an HVAC drawing. In a forced-air system, fans move the heated or cooled air through ductwork into the working areas, offices, and public spaces. Air return ducts collect air from these areas and channel it back through the system or to the outside atmosphere.

The third drawing of *Figure 26* is a section of a typical shop drawing showing ductwork details for a project. The ductwork for these systems is usually one of the following types:

- Individual duct system
- Trunk duct system
- Crawl space plenum or plenum ceiling systems

The first two delivery systems may be used either from the floor or ceiling levels. The ductwork begins at the central unit and moves out to each area of work space. The plans may specify circular or rectangular ducts. Circular ducts slip together and bend to form angles. Rectangular ducts fit between joists and studs. In the individual duct system, many small ducts may reach from the central unit to each room, where a grilled register delivers the air. In a trunk duct system, large main ducts run the length of the structure and smaller ducts branch off to each room register.

In some forced-air systems, particularly in buildings with a crawl space, the heating unit is designed to heat the exterior walls and the floor above the crawl space. A crawl space plenum system heats the entire crawl space and the conditioned air rises through floor registers. This system requires no ductwork to transfer the warm air since it is transferred through the structural frame.

To gain the most information from HVAC plans, you should look for the following:

- The direction of the central fan
- Duct supply lines, which deliver the conditioned air to all parts of the structure
- Return duct (some HVAC systems have return ducts to return air to the central unit for more heating or cooling)

3.3.3 Sprinkler Systems

Sprinkler contractors use the architectural and structural drawings of the project to clarify the requirements for any sprinkler system installation. The basic sprinkler system drawings for a building are plan view drawings that contain most of the information for a sprinkler system layout. Reflected ceiling plans typically show tile layouts, light fixture locations, sprinkler head locations, HVAC diffuser locations, and so on. From this information, drawings are developed to show the detailed data necessary for an installation. *Figure 27* is a typical sprinkler shop drawing showing the detailed sprinkler system layout for a warehouse area and an office area.

3.4.0 Electrical Drawings

Electrical drawings identify the layout of the electrical distribution system, the lighting requirements, and the telecommunications and computer connections. This information is provided in plan view on a copy of the floor plan.

Although you do not need to understand the details of electrical circuitry to read an electrical plan, you should understand the basic design of the system. All systems begin at the service source. Like water and natural gas, the process begins with a meter being installed by the utility company. From the meter, service moves through a main cutoff switch to the service entry panel. The panel separates service into branch circuits. These circuits are protected from overload by circuit breakers with various capacities. The branches then feed out to specific fixtures and outlets throughout the building.

Electrical systems, like plumbing systems, may be very basic for structures such as general warehouses and office buildings. They can be very complicated for research laboratories and hospitals. Some complex facilities have backup generators and parallel wiring systems.

3.4.1 Power Plan

A typical power plan illustrates the power requirements of the structure. *Figure 28* is a riser diagram and would normally show the panels, receptacles, and the circuitry of power-utilizing equipment. Some buildings need multiple panels. The power plan would then include panel schedules. Panel schedules list the circuits in the panel, the individual power required for each panel, and the total power requirement for the system. The schedules total the power requirements to help the electrical contractor size the panel. The power company sizes the overall service requirements.

3.4.2 Lighting Plan

A typical lighting plan locates the various lighting fixtures in the building (*Figure 29*). It is complemented by a fixture schedule that lists the types of light fixtures to be used. The fixture schedule is organized by number or letter. These refer to the manufacturer and model, the wattage of the lamps, voltage, and any special remarks concerning the fixture. Outlets vary according to their electrical capacity and type. Different symbols are used for outlets according to their voltage requirements. This is also true of most other types of electrical items. Details such as type, size, and voltage are listed on the schedules.

The lighting plan can also include smoke- and fire-detection equipment, emergency lighting, cable TV, and telephone outlets. Buildings with more complex electrical systems may include separate plans for fire prevention and telecommunications systems.

201F27.EPS

Figure 27 ♦ Typical sprinkler shop drawing.

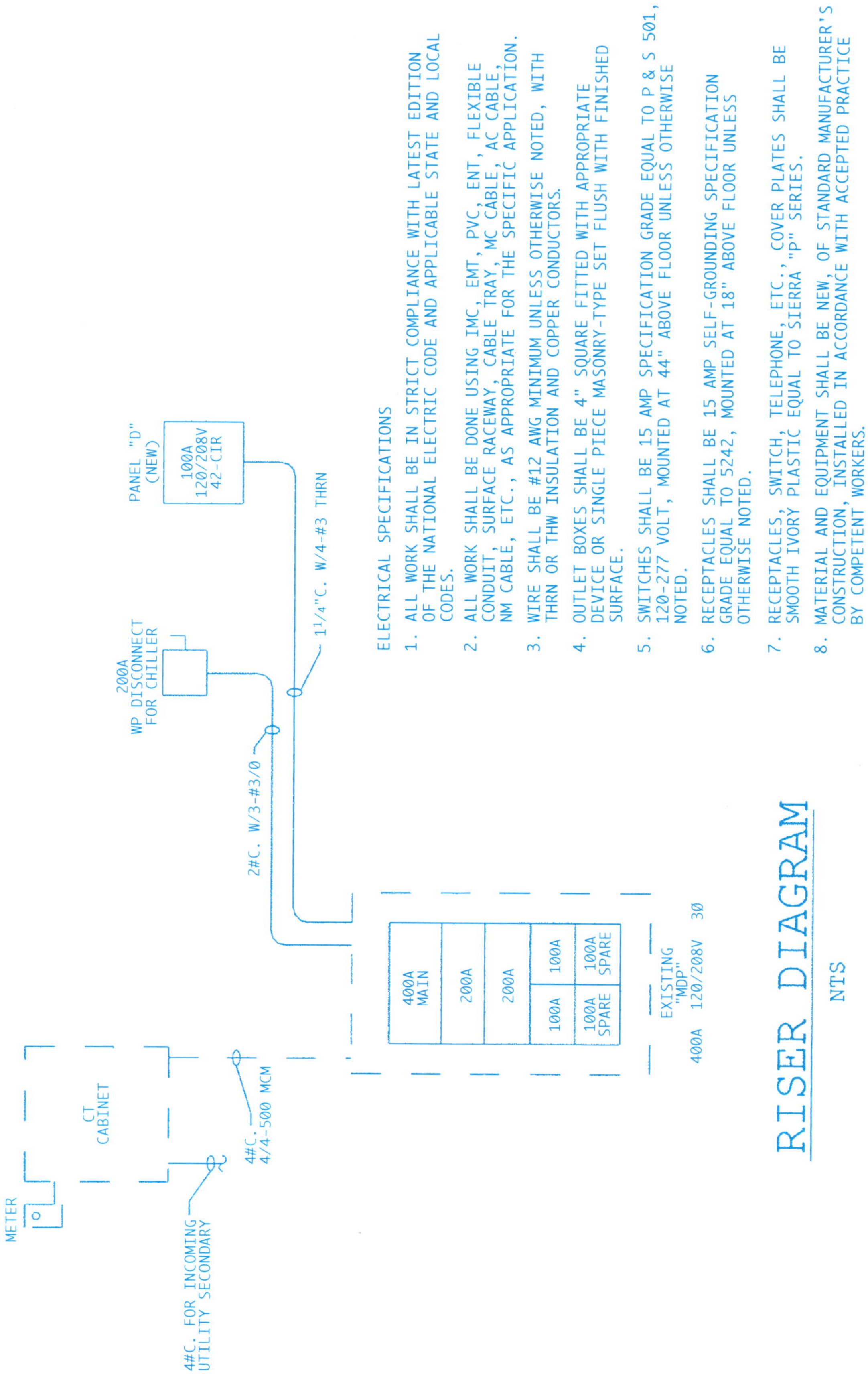

201F28.EPS

Figure 28 ◆ A power plan.

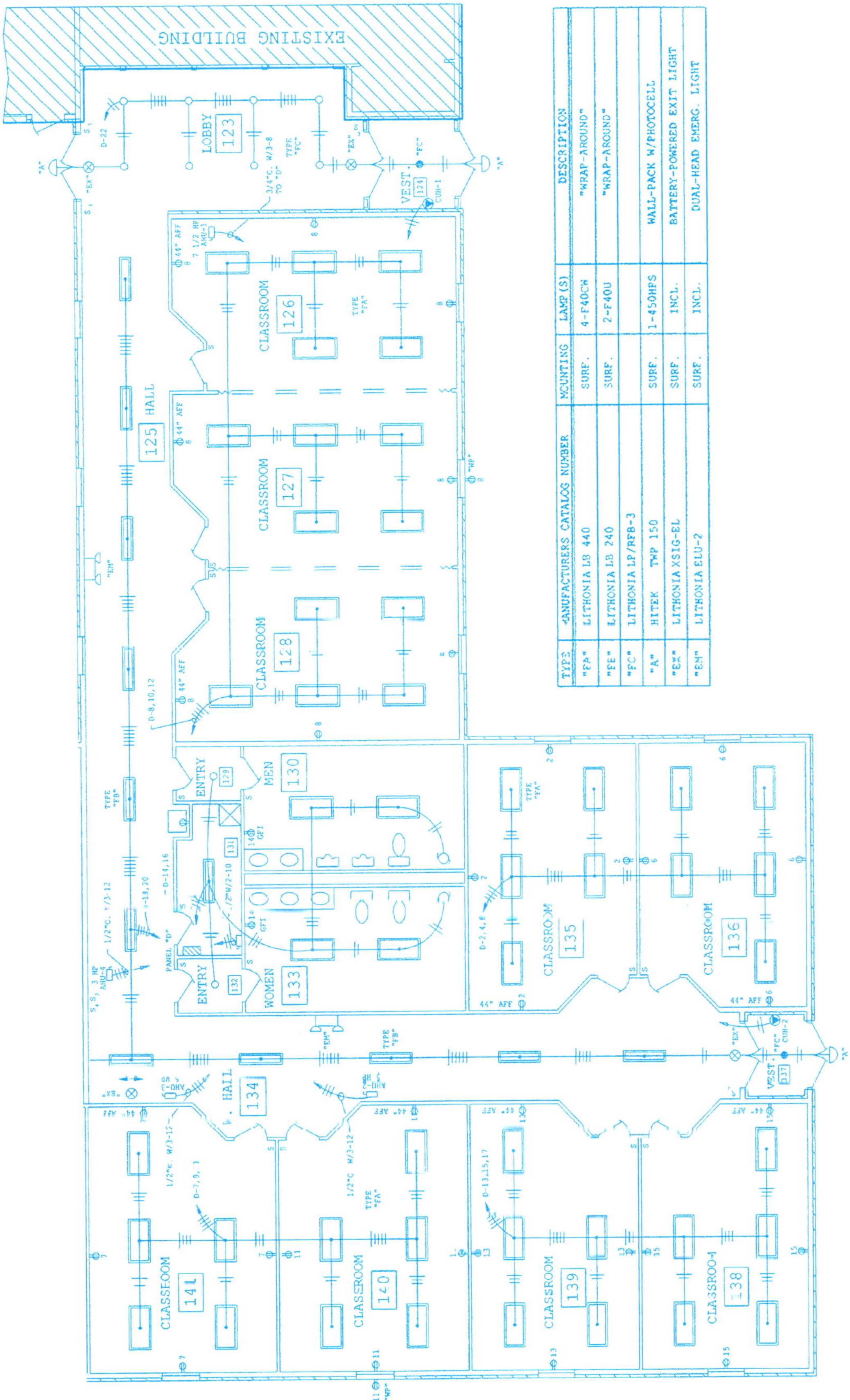

TYPE	MANUFACTURERS CATALOG NUMBER	MOUNTING	LAMP(S)	DESCRIPTION
"FA"	LITHONIA LB 440	SURF.	4-F40CW	"WRAP-AROUND"
"FB"	LITHONIA LB 240	SURF.	2-F40U	"WRAP-AROUND"
"FC"	LITHONIA LP/RFB-3			
"A"	HITEK TWP 150	SURF.	1-450HPS	WALL-PACK W/PHOTOCELL
"EX"	LITHONIA XSIG-EL	SURF.	INCL.	BATTERY-POWERED EXIT LIGHT
"EM"	LITHONIA ELU-2	SURF.	INCL.	DUAL-HEAD EMERG. LIGHT

Figure 29 ◆ A lighting plan.

DID YOU KNOW?

Lighting the New York Skyline

The first searchlight on top of the Empire State building heralded the election of Franklin D. Roosevelt in 1932. A series of floodlights were installed in 1964 to illuminate the top 30 floors of the building. Today, the color of the lights is changed to mark various events. Yellow marks the US Open, and red, white, and blue mark Independence Day.

The building is lit from the 72nd floor to the base of the TV antenna by 204 metal halide lamps and 310 fluorescent lamps. In 1984, a color-changing apparatus was added in the uppermost mooring mast. There are 880 vertical and 220 horizontal fluorescent lights. The colors can be changed with the flick of a switch.

On Earth Day and Rainforest Awareness Day, the Empire State Building's tower lights are green.

4.0.0 ♦ WRITTEN SPECIFICATIONS

The written specifications for a building or project are the descriptions in writing of work and duties required of the owner, architect, and consulting engineer. Together with the working drawings, these specifications form the basis of the contract requirements for the construction of the building or project. Those who use the construction drawings and specifications must always be alert to discrepancies between the working drawings and the written specifications. The following are some situations where discrepancies may occur:

- Architects or engineers use standard or prototype specifications and attempt to apply them without any modification to specific working drawings.
- Previously prepared standard drawings are changed or amended by reference in the specifications only, and the drawings themselves are not changed.
- Items are duplicated in both the drawings and specifications, but an item is subsequently amended in one and overlooked in the other contract document.

Legally, specifications take precedence over drawings. However, the person in charge of the project has the responsibility to determine whether the drawings or the specifications take precedence in a given situation.

4.1.0 How Specifications are Written

Writing accurate and complete specifications for building construction is a serious responsibility for those who design the buildings. This is because the specifications, combined with the working drawings, govern practically all important decisions made during the construction span of every project. Compiling and writing these specifications is not a simple task, even for those who have considerable experience in preparing such documents.

A set of written specifications for a single project usually contains thousands of products, parts, and components, and the methods of installing them, all of which must be covered in either the drawings and/or specifications. No one can memorize all of the necessary items required to describe accurately the various areas of construction on a project. The person writing the specifications must rely upon reference materials such as manufacturer's data, catalogs, and checklists to produce high-quality specifications.

4.2.0 Format of Specifications

The most suitable organization of the specifications is a series of sections dealing with the construction requirements, products, and activities. Specifications should be easily understandable by the different trades. Those who use them must be able to find all information needed without spending too much time looking for it.

The most commonly used specification format in North America is *MasterFormat*™. This standard was developed jointly by the Construction Specifications Institute (CSI) and Construction Specifications Canada (CSC). For many years prior to 2004, the organization of construction specifications and supplier's catalogs was based on a standard with 16 sections, known as divisions. The divisions and their subsections were individ-

INSIDE TRACK

Specifications

Written specifications supplement the related working drawings and contain details not shown on the drawings. Specifications define and clarify the scope of the job. They describe the specific types and characteristics of the components that are to be used on the job and the methods for installing some of them. Many components are identified specifically by the manufacturer's model and part numbers. This type of information is used to purchase the various items of hardware needed to accomplish the installation in accordance with the contractual requirements.

ually identified by a five-digit numbering system. The first two digits represented the division number and the next three individual numbers indicated successively lower levels of classification. For example, the number 13213 represented division 13, subsection 2, sub-subsection 1 and sub-sub-subsection 3.

In 2004, the *MasterFormat*™ standard underwent a major change. What had been 16 divisions was expanded to four major groupings and 49 divisions with some divisions reserved for future expansion (*Figure 30*). The first 14 divisions are essentially the same as the old format. Subjects under the old Division 15 – *Mechanical* have been relocated to new divisions 22 and 23. The basic subjects under old Division 16 – *Electrical* have been relocated to new divisions 26 and 27.

In addition, the numbering system was changed to 6 digits to allow for more subsections in each division, for finer definition. In the new numbering system, the first two digits represent the division number. The next two digits indicate subsections of the division, and the two remaining digits signify the third level sub-subsection numbers. The fourth level, if required, is a decimal and number added to the end of the last two digits. For example, the number 132013.04 indicates division 13, subsection 20, sub-subsection 13, and sub-sub-subsection 04.

Division Numbers and Titles

Procurement and Contracting Requirements Group

Division 00 **Procurement and Contracting Requirements**

Specifications Group

General Requirements Subgroup

Division 01 **General Requirements**

Facility Construction Subgroup

Division 02 **Existing Conditions**
Division 03 **Concrete**
Division 04 **Masonry**
Division 05 **Metals**
Division 06 **Wood, Plastics, and Composites**
Division 07 **Thermal and Moisture Protection**
Division 08 **Openings**
Division 09 **Finishes**
Division 10 **Specialties**
Division 11 **Equipment**
Division 12 **Furnishings**
Division 13 **Special Construction**
Division 14 **Conveying Equipment**
Division 15 *Reserved*
Division 16 *Reserved*
Division 17 *Reserved*
Division 18 *Reserved*
Division 19 *Reserved*

Facility Services Subgroup

Division 20 *Reserved*
Division 21 **Fire Suppression**
Division 22 **Plumbing**
Division 23 **Heating, Ventilating, and Air Conditioning**
Division 24 *Reserved*
Division 25 **Integrated Automation**
Division 26 **Electrical**
Division 27 **Communications**
Division 28 **Electronic Safety and Security**
Division 29 *Reserved*

Site and Infrastructure Subgroup

Division 30 *Reserved*
Division 31 **Earthwork**
Division 32 **Exterior Improvements**
Division 33 **Utilities**
Division 34 **Transportation**
Division 35 **Waterway and Marine Construction**
Division 36 *Reserved*
Division 37 *Reserved*
Division 38 *Reserved*
Division 39 *Reserved*

Process Equipment Subgroup

Division 40 **Process Integration**
Division 41 **Material Processing and Handling Equipment**
Division 42 **Process Heating, Cooling, and Drying Equipment**
Division 43 **Process Gas and Liquid Handling, Purification, and Storage Equipment**
Division 44 **Pollution Control Equipment**
Division 45 **Industry-Specific Manufacturing Equipment**
Division 46 *Reserved*
Division 47 *Reserved*
Division 48 **Electrical Power Generation**
Division 49 *Reserved*

201F30.EPS

Figure 30 ◆ 2004 MasterFormat™.

Review Questions

1. In some commercial drawing sets, the site plans are called _____ drawings.
   a. civil
   b. commercial
   c. layout
   d. elevation

2. Architectural drawings are identified with the letter _____.
   a. E
   b. A
   c. M
   d. P

3. Structural drawings usually include _____ plans.
   a. HVAC
   b. riser
   c. piping
   d. foundation

4. A second-floor structural drawing would show the _____.
   a. placement of partition walls
   b. placement of electrical service drops
   c. floor layout
   d. configuration of loadbearing members

5. Mechanical drawings for HVAC systems are typically shown in a(n) _____ view.
   a. section
   b. plan
   c. elevation
   d. cutaway

6. The emergency lighting systems are part of the _____ drawings.
   a. electrical
   b. civil
   c. mechanical
   d. architectural

7. When reading commercial plans, use a _____ to avoid missing details.
   a. step-by-step process
   b. schedule
   c. local code
   d. plan handbook

8. Topographical information on a site plan includes _____.
   a. street names
   b. zoning requirements
   c. roofing materials
   d. lot corner elevations

9. On a commercial site plan, only existing contour lines are shown.
   a. True
   b. False

10. A feature on site plans that may be found only on commercial plans are _____.
    a. directional arrows
    b. exterior wall positioning data
    c. landscaping features
    d. grading details

11. To help clarify ambiguous parts of the drawing, _____ may be written on commercial floor plans.
    a. notes
    b. details
    c. symbols
    d. schedules

12. Of the following, the _____ would be seen in a door schedule.
    a. door size
    b. riser diagram
    c. room location
    d. delivery date

13. An enlargement of an area on the floor plan is called a(n) _____.
    a. elevation
    b. schedule
    c. section
    d. detail

14. Interior elevation drawings may have a scale as small as _____ inch to 1 foot, or as large as _____ to 1 foot.
    a. ¼; ¾
    b. ⅛; ¼
    c. ⅜; ⅝
    d. ½; ¾

Review Questions

15. Stairways will usually be shown as elevations and _____ on the drawings.
    a. sections
    b. grids
    c. materials
    d. schedules

16. Floor framing plans are found in the _____ drawings.
    a. architectural
    b. structural
    c. mechanical
    d. electrical

17. The placement of a column in a structural drawing will be shown by a(n) _____.
    a. circle with a smaller circle inside it
    b. circle with an X through it
    c. callout sequence number or mark
    d. an open grid square

18. The structural engineer draws a framing plan for _____.
    a. each floor and the roof
    b. the plumbing, electrical, and HVAC systems
    c. landscaping and grading
    d. the foundation and exterior walls

19. The arrangement of beams, joists, and girders is shown on _____.
    a. elevations
    b. sections
    c. schedules
    d. framing plans

20. It is easier to read framing plans by _____.
    a. starting at the bottom and working upwards
    b. starting at the top and working downwards
    c. isolating the separate bays or spans between columns
    d. reading the foundation plans first

21. Concrete is used in building foundations because it has _____.
    a. low compressive strength
    b. high compressive strength
    c. low tensile strength
    d. high tensile strength

22. Steel reinforcement (rebar) is used in concrete foundations to increase the _____ strength of concrete.
    a. low compressive
    b. high compressive
    c. low tensile
    d. high tensile

23. For bonded post-tensioning systems used in elevated floors, the tendons are _____.
    a. threaded through plastic sleeves and tensioned after the concrete has cured
    b. tensioned before the concrete has been placed
    c. embedded in the concrete when the concrete is placed
    d. embedded in the concrete and tensioned off site at a manufacturing facility

24. A plumbing riser diagram shows _____.
    a. vertical pipes in the wall
    b. horizontal distance or pipe runs
    c. the hot- and cold- water supply lines
    d. the couplings, elbows, tees, and valves needed

25. Piping for distribution of natural gas is included in the _____ plans.
    a. HVAC
    b. plumbing
    c. structural
    d. mechanical

26. Mechanical plans typically contain schedules that identify the different types of _____ equipment.
    a. heavy
    b. grading
    c. safety
    d. HVAC

Review Questions

27. The electrical system begins with a _____.
    a. service entry panel
    b. main cut-off switch
    c. circuit breaker
    d. meter

28. There may be discrepancies between the written specifications and the working drawings because _____.
    a. the specifications were amended and the drawings were not
    b. they refer to different parts of the project
    c. specifications are only used by the engineer and architect
    d. specifications are no longer needed once the drawings are prepared

29. Specifications follow a standard format _____.
    a. as outlined in the building codes
    b. for convenience in writing and ease of reference
    c. because it is required by law
    d. specified by ANSI

30. The most commonly used specification is developed by the _____ and Construction Specifications Canada (CSC).
    a. Brick Industry Association (BIA)
    b. American National Standards Institute (ANSI)
    c. Construction Specifications Institute (CSI)
    d. International Code Council (ICC)

Summary

A typical commercial plan set includes architectural, structural, mechanical, plumbing, and electrical drawings with a letter designation for each section. Each section may include more than one sheet, depending on the size and complexity of the project. Architectural drawings include the site plan, floor plans, wall sections, door and window details, and schedules. Structural drawings contain foundation plans, framing plans, support details, schedules, and notes. Mechanical drawings include HVAC plans in both plan view and isometric view. Plumbing drawings and mechanical drawings are similar in format. They provide plan and isometric views of the fresh water and waste water systems. Electrical drawings usually include a power plan and a lighting plan, which may be on the same or separate sheets, depending on the complexity of the project.

When reading commercial plans, follow a step-by-step process to avoid confusion and take note of all the details involved. Begin by reading the project specifications to pick up details not found on the drawings. Quickly review all the drawings to get a general impression of the shape, size, and appearance of the structure. Begin correlating the floor plans with the exterior elevations. Look at the wall sections to determine the wall types and construction details.

Review the structural plans to determine the foundation requirements and the type of structural system. Carefully consider all items that may require special construction procedures. Review all interior elevations. Try to get a clear picture of what the interior of the building will look like. Review the finish schedule and the mechanical and electrical plans. In particular, look for work items to be performed by drywall technicians or features that will affect their work. If some part of the drawing is not clear, check with your supervisor. There may be a logical explanation, or there may be a mistake. The architect should clarify and resolve all conflicts in writing.

A set of written specifications for a large commercial project usually contains thousands of products, parts, components, and the methods of installing them. A standard format has been voluntarily adopted for commercial construction throughout North America. Manufacturers and suppliers key their catalogs and data to this standard format.

Notes

Trade Terms Introduced in This Module

ADA Requirements: The Americans with Disabilities Act of 1990 requires the use of accessible design methods for the construction of public accommodations such as lodging, public parks, schools, and restaurants.

As-built: Revised drawings of an installation that show the changes made during installation.

Beams: Loadbearing horizontal framing elements supported by walls or columns and girders.

Building codes: Codes published by state and local governments to establish minimum standards for various types of interior and exterior construction.

Callouts: Markings or identifying tags describing parts of a drawing; callouts may refer to detail drawings, schedules, or other drawings.

Civil drawings: A drawing that shows the overall shape of the building site. It is also called a site plan.

Compressive strength: Refers to a material's ability to support weight.

Contour lines: Imaginary lines on a site plan/ plot plan that connect points of the same elevation. Contour lines never cross each other.

Elevation view: A drawing giving a view from the front, rear, or side of a structure.

Girders: Large steel or wooden beams supporting a building, usually around the perimeter.

Isometric drawing: A three-dimensional type of drawing in which the object is tilted so that all three faces are equally inclined to the picture plane.

Joists: Horizontal members of wood or steel supported by beams and holding up the planks of floors or the lathes of ceilings. Joists are laid edgewise to form the floor support.

Landscape drawings: A drawing that shows proposed plantings and other landscape features.

Plan view: A drawing that represents a view looking down on an object.

Riser diagram: A type of isometric drawing that depicts the layout, components, and connections of a piping system.

Sections: Drawings that detail construction techniques or materials.

Schedules: Charts or tables that provide detailed information corresponding to various parts of the drawings.

Tensile strength: Refers to the flexibility of a material.

Title block: A section of an engineering drawing blocked off for pertinent information, such as the title, drawing number, date, scale, material, draftsperson, and tolerances.

Additional Resources and References

Additional Resources

This module is intended to be a thorough resource for task training. The following reference works are suggested for further study. These are optional materials for continued education rather than for task training.

Architectural Graphic Standards, Eighth Edition, The American Institute of Architects, John Wiley & Sons, New York, NY, 1988.

Basics for Builders: Plan Reading & Material Takeoff, Wayne J. DelPico, R.S. Means Company, Inc., Kingston, MA, 1994.

Figure Credits

Topaz Publications, Inc., 201F01 (top), 201F22, 201F23

Benson Group Companies, 201F01 (bottom)

Dumond Chemicals, Inc., 201SA01

Marek Brothers Systems, Inc., 201F05, 201F24

Tom Crane, Photographer, The Athenaeum of Philadelphia, 201SA02

Ivey Mechanical Company, LLC, 201F15, 201F16

Kogok Corporation, 201F26C

Kaufman Fire Protection, 201F27

The Construction Specifications Institute (CSI) and Construction Specifications Canada (CSC), 201F30

The Numbers and Titles used in this textbook are from *MasterFormat*™, published by The Construction Specifications Institute (CSI) and Construction Specifications Canada (CSC), and are used with permission from CSI. For those interested in a more in-depth explanation of *MasterFormat*™ and its use in the construction industry visit www.csinet.org/masterformat or contact:

The Construction Specifications Institute (CSI)
99 Canal Center Plaza, Suite 300
Alexandria, VA 22314
800-689-2900; 703-684-0300

NCCER CURRICULA — USER UPDATE

NCCER makes every effort to keep its textbooks up-to-date and free of technical errors. We appreciate your help in this process. If you find an error, a typographical mistake, or an inaccuracy in NCCER's curricula, please fill out this form (or a photocopy), or complete the online form at **www.nccer.org/olf**. Be sure to include the exact module ID number, page number, a detailed description, and your recommended correction. Your input will be brought to the attention of the Authoring Team. Thank you for your assistance.

Instructors – If you have an idea for improving this textbook, or have found that additional materials were necessary to teach this module effectively, please let us know so that we may present your suggestions to the Authoring Team.

NCCER Product Development and Revision

13614 Progress Blvd., Alachua, FL 32615

Email: curriculum@nccer.org

Online: www.nccer.org/olf

❑ Trainee Guide ❑ Lesson Plans ❑ Exam ❑ PowerPoints Other ______________________

Craft / Level: Copyright Date:

Module ID Number / Title:

Section Number(s):

Description:

Recommended Correction:

Your Name:

Address:

Email: Phone:

45202-09

Steel Framing

45202-09
Steel Framing

Topics to be presented in this module include:

Overview

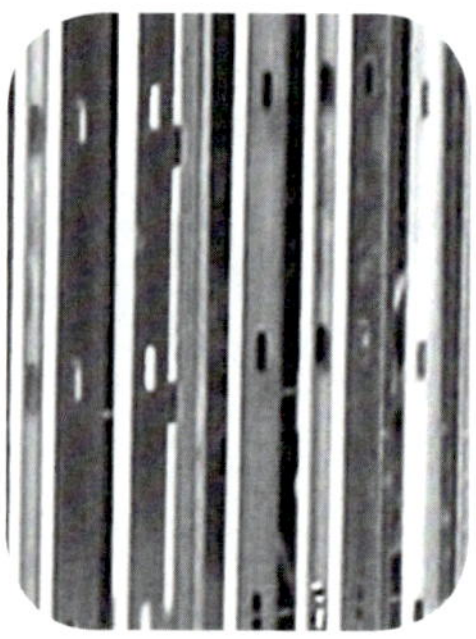

In commercial and multi-family residential construction, it is common to use steel framing materials in place of wood studs to frame walls and partitions. It is also becoming more common in single-family residential construction as the price of lumber rises. This module describes the uses and installation of cold-formed steel framing.

As the demand for more cost-efficient, more sustainable construction grows, builders and developers are specifying both structural and nonstructural applications of cold-formed steel. Recognition and acceptance of cold-formed steel in building codes is leading to a better understanding of its capabilities and benefits.

Low and mid-rise structures using cold-formed steel as the main structural component have been successfully built throughout the United States, creating more durable structures and lowering construction costs. Steel framing, however, requires a worker to master tools and joining techniques different from those required in wood framing construction. To work a variety of steel framing projects, from installing loadbearing steel assemblies to nonstructural walls, you will need to become familiar with these materials and methods.

Objectives

When you have completed this module, you will be able to do the following:

1. Identify the components of a steel framing system.
2. Identify and select the tools and fasteners used in a steel framing system.
3. Identify applications for steel framing systems.
4. Build back-to-back, box, and L-headers.
5. Lay out and install a steel stud structural wall with openings to include bracing and blocking.
6. Lay out and install a steel stud nonstructural wall with openings to include bracing and blocking.
7. Interpret engineer drawings/shop installations.
8. Follow shop drawings to build:
   - Windows/openings
   - Bottom track
   - Top track
   - Bypass clips
   - Deflection cups
9. Integrate work of other trades:
   - Window
   - Water management
   - Stone/brick
   - Miscellaneous iron support

Trade Terms

Blocking
Clip angle
C-shape
Cold-formed steel
Curtain wall
Diaphragm
For-fill
Header
Knurled
Lateral
Lips
Mil
Panelization
Plenum
Racking
Rim track
Shear wall
Spandrel
Track

Prerequisites

Before you begin this module, it is recommended that you successfully complete *Core Curriculum*; *Drywall Level One*; and *Drywall Level Two*, Module 45201-09.

This course map shows all of the modules in the second level of the *Drywall* curriculum. The suggested training order begins at the bottom and proceeds up. Skill levels increase as you advance on the course map. The local Training Program Sponsor may adjust the training order.

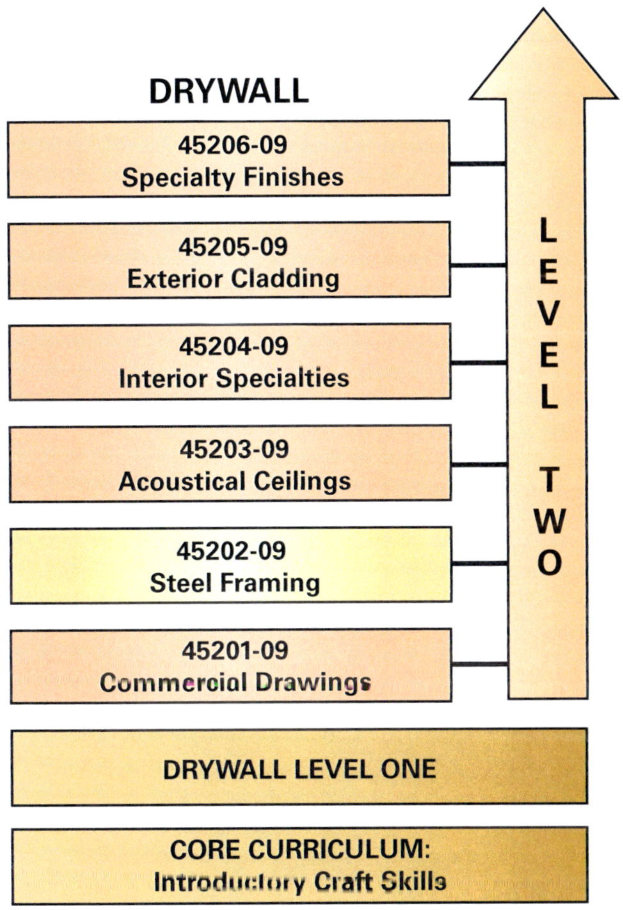

202CMAP.EPS

1.0.0 ♦ INTRODUCTION

Most commercial construction uses light steel framing for exterior and interior walls. Although most residential construction now uses wood framing, the use of light steel is on the increase. Designers and builders know that steel is strong, durable, and functional. Steel framing may be used inside a structure for room partitions and for loadbearing or nonbearing walls. Some of the advantages of using steel framing instead of wood include the following:

- Steel studs and joists are strong, lightweight, and made from uniform material that will not warp, crack, rot, or split. Steel has the highest strength-to-weight ratio of any building material. Also, steel framing members are made with consistent dimensions, so assembly is faster and easier than with wood products.
- The exterior of steel members is uniform and provides a flat surface for sheathing or other material attachment. Steel is compatible with all sheathing materials.
- Steel framing does not contribute combustible material to feed a fire. Its lower combustion rating means lower insurance costs for builders and owners.
- Steel can be engineered to meet the toughest wind and seismic ratings specified by building codes.

In addition to its good structural features, steel has some important environmental qualities. Industry-wide, steel has an average of 67 percent recycled content, making it the world's most recycled material. In the United States alone, almost 70 million tons of steel per year are recycled or exported for recycling. **Cold-formed steel** framing supports industry efforts to promote sustainable construction.

> **NOTE**
> *Appendix A* contains a list of terms commonly used in conjunction with cold-formed steel framing.

By comparison, wood framing is heavy and more difficult to move. Wood comes from trees, which contain a lot of water. As the wood dries on the job site, it shrinks and changes shape. This can cause a great deal of waste when building with wood products. After construction, the wood drying process causes a building to settle. The settling process can lead to cracked drywall, cracked and uneven floors and ceilings, poorly fitting doors and windows, and nail pops.

1.1.0 Structural Engineering

A great deal of thought goes into the design of a building. As you read the following section, you may conclude that structural engineering of a building is a fairly easy task. While on the surface, structural design may appear to be the product of simple common sense, it actually requires a thorough knowledge of math and physics. This is why a qualified engineer must review and approve all construction plans. Even changes made in the field need to be recorded, and then reviewed and approved by an engineer.

Structures are designed to withstand the loads to which they will be subjected. When a building has too little structural support, it can collapse under stress. When a building has too much structural support, it wastes building materials, labor, and time. Designers study the internal and external stressors to which a building will be subjected, and then they design the structure to withstand them. As a drywall technician, your job will be to follow the instructions given on the engineering and shop drawings and in the project specifications.

In this section, the following building stressors (*Figure 1*) will be covered:

- *Dead loads* – Refers to the weight of the building materials on the structure. This includes the weight of the roof and any heavy equipment, such as air conditioning towers and swamp coolers, installed on the roof. If a crane is used on the roof during the construction of the building, it is also considered a dead load.

INSIDE TRACK

Wood-Framed Construction

Wood-framed construction places heavy demands on timber resources. It takes the wood from roughly one acre of forest to build a 2,000 square-foot home.

North American Codes and Standards

INSIDE TRACK

The use of cold-formed steel (CFS) members in building construction began around 1850. In North America, however, steel members were not widely used until 1946, when the American Iron and Steel Institute (AISI) Specification was first published. This design standard was primarily based on research sponsored by AISI at Cornell University. Subsequent revisions to the document reflected technical developments and ultimately led to the publishing of the *North American Specification for the Design of Cold-Formed Steel Structural Members 2001 Edition*. AISI, along with the American National Standards Institute (ANSI), American Society of Testing and Materials International (ASTM), and the International Codes Council (ICC) Steel Stud Manufacturers Association (SSMA) govern the design, manufacture, and use of cold-formed steel and framing. All framing members carry a product identification to comply with the minimum sheet steel thickness, the coating designation, minimum yield strength and the manufacturer's name.

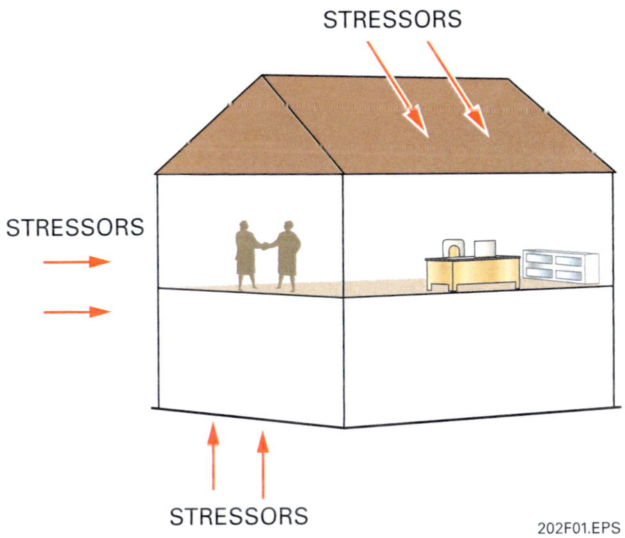

Figure 1 ◆ Building stressors.

- *Live loads* – Result from using the building when it is finished and occupied. Examples of live loads include people moving about in the building as they work and the equipment that is transported through the building.
- *Other loads* – In addition to dead and live loads, buildings are subjected to stress from wind, rain, snow, and in some areas, earthquakes.

It is sometimes helpful to use pictures to explain the various types of loads. In the first example of a dead load (*Figure 2*), a person is standing very still in the middle of a board that is securely nailed between two buildings. The board is thick and wide enough to support the person's weight, and the distance between the buildings is short, so the board barely sags at all.

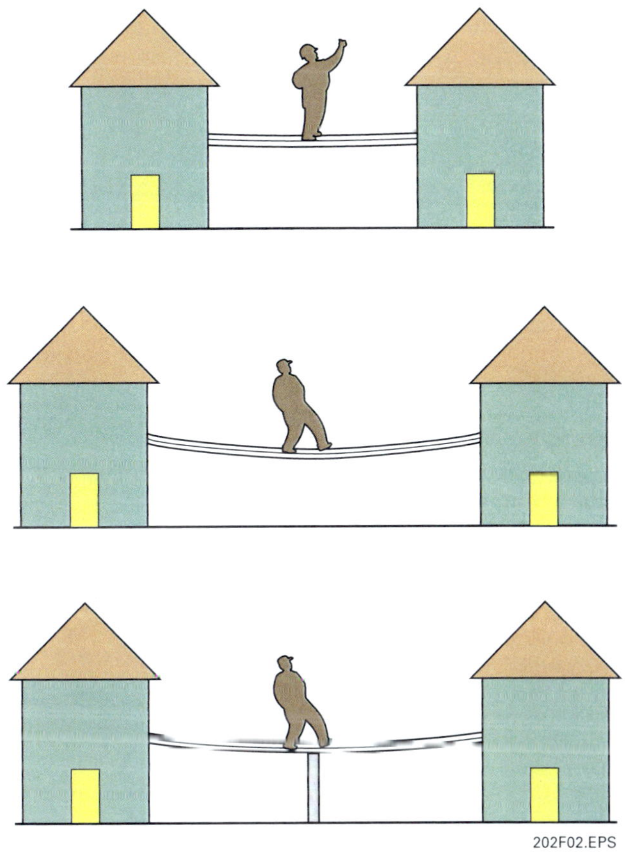

Figure 2 ◆ Types of loads.

In the second example of a dead load in *Figure 2*, the person is still standing in the middle of the board, but the distance between the buildings and the length of the board have doubled. The board is now sagging in the middle, causing the person to feel insecure. This sagging is called vertical deflection. It results from the person's weight applying downward pressure on the board. If the person began jumping up and down in the middle of the board, causing it to sag even more, this would be an example of a live load. If the person continued jumping, the board could even break. The examples of loads just described are the same types that a typical building must be able to safely support.

What can be done to decrease the dead load on the board and the sagging (vertical deflection) resulting from the person's weight? One solution would be to place a support under the middle of the board as shown in the third example of *Figure 2*. This would add structural support to the board and help to sustain it by transferring any weight load on the board to the ground. This is the basic idea behind structural engineering.

One type of force that is of special importance to tall structures is wind. Wind causes tall buildings to move in a manner similar to the way a tall tree sways in the wind. This movement is called horizontal deflection, also known as **lateral** drift. All tall buildings are designed to move with the wind. A wind blowing at 110 miles an hour will move the Empire State Building about 1.48 inches at the top.

Designers must address all the various types of stressors when designing buildings. Whether the building is constructed of wood framing or steel framing, the structural factors are the same. The building must support its dead load and its live load, as well as other loads, such as snow, rain, and wind.

Buildings are constructed so that the weight from all loads is transferred through structural members to the foundation until all of the weight is transferred to the ground. This design transfers the weight of the roof to the wall posts, then to the floors and foundation. Finally, all the weight of the building is transferred to the ground.

Not all wall studs are loadbearing. Some walls only need to support their own weight. Interior and exterior walls may be either loadbearing or nonbearing. Loadbearing walls need to be strong, and they require a great deal of support. Nonbearing walls only need to support themselves. **Curtain walls** are exterior walls that are designed to support their own weight and to resist horizontal deflection, but they do not carry any other load.

Because door and window openings cause a break in the wall stud pattern, they need to be reinforced on the top and on each side to make up for the missing studs. **Headers** are strong structural members that are placed horizontally over the door and window openings. Trimmer studs that support the headers are placed on each side of the opening. Beneath the window openings, cripple studs are used to support the sill. Window and door openings may be purchased as premanufactured items, or they may be built on the job site. To increase structural supports, trimmer studs and end cripple studs are butted against and secured to the full-length wall studs, called king studs. Steel framing members may be fastened together by screws, welding, or cinching according to the project specifications.

2.0.0 ◆ TOOLS FOR STEEL FRAMING WORK

Metal stud work requires the use of more power tools than wood frame assemblies. Only use power tools that you have been trained to use.

CAUTION

Always practice power tool safety. This includes wearing all of the appropriate personal protective equipment and ensuring that all electrical tools are grounded.

Some of the common tools that are used in steel framing component assembly include the following:

- *Screwgun* – This tool (*Figure 3*) is used to connect steel members together and to attach sheathing material, such as plywood and gypsum board, to steel. A screwgun is also called a power screwdriver. It increases the speed and efficiency of construction work. The preferred screwgun for steel-to-steel connections should have an adjustable clutch and torque setting, with a speed range of from 0 to 2,500 rpm. A drywall screwgun (0 to 4,000 rpm) is recommended for attaching plywood or gypsum board to steel. In addition to standard and cordless screwguns, there is a version that is powered by compressed air. Feed attachments are also available that automatically supply screws into the gun (*Figure 4*).

CAUTION

Never use a drywall screw gun for steel-to-steel connections. If the screwgun is running too fast, the tip of the screw may burn out before it penetrates the steel. Once the screw has been properly seated, the screwgun will automatically stop spinning the screw, preventing it from stripping. The high-impact gun is designed to prevent this problem.

202F03.EPS

Figure 3 ◆ Cordless screwgun.

202F04.EPS

Figure 4 ◆ Screwgun with collated feed.

NOTE

A power screwdriver is not just a drill with a screwdriver bit. It normally has an adjustable depth control to prevent overdriving the screws. Many have a clutch mechanism that disengages when the screw has been driven to a preset depth. Some power screwdrivers are designed to perform specific fastening jobs, such as fastening drywall to walls and ceilings.

- *Stud driver* – Two types of stud drivers are the powder-actuated fastener gun (*Figure 5*) and the hammer drill (*Figure 6*). They are used to drive fasteners into concrete slabs, structural steel, and foundations. Operators must be properly trained in the use of these tools and must have an operator's certificate before using them.

202F05.EPS

Figure 5 ◆ Powder-actuated installation tool.

202F06.EPS

Figure 6 ◆ Hammer drill.

WARNING!

Stud drivers are high-powered tools that can cause serious injuries if used improperly. Always wear a hard hat, heavy boots, and eye, ear, and face protection when using these tools. Treat a powder-actuated stud driver as if it were a loaded gun. Before you handle it, determine whether it is loaded with a powder charge and a fastener—either can be dangerous. Always point the tool away from yourself and others. Make sure that you know when the tool is in the locked and unlocked positions.

Powder-actuated fasteners can be used only by personnel trained and certified in the use of the specific tool. You must carry your certification card with you when using the tool, and the card must be valid for the tool you are using.

Be sure to select the proper charge for the job in accordance with the manufacturer's instructions. If in doubt, check with your supervisor.

Many factors affect the holding strength of fasteners in concrete. These factors include the compressive strength of the concrete, the fastener shank diameter, and how deep the fastener penetrates into the concrete. Fastener holding strength is also determined by the spacing between fasteners and the condition of the edge of the concrete. Headed or threaded drive pins are available with **knurled** or spiraled shanks to increase holding power in structural steel material. Fasteners inserted under tension may require washers to prevent them from pulling through the steel **track**.

- *Locking C-clamps and bar clamps* – Clamps are used to hold steel members securely together during fastening, which is very important in steel framing. This is because as the screw penetrates the first layer of steel, it tends to pull that layer away from the layer to which it is being attached. This tendency is called climbing the threads. When it happens, it leaves a gap between the members that can decrease the strength of the joint. To prevent this, clamp the members together with the clamp tips positioned as close as possible to the fastening point. Use as many clamps as are needed to secure the components. All clamps used in steel framing should have regular tips on the ends, without fixed or swivel pads. This allows all of the pressure to be placed on a small area instead of being dispersed across the pads.

  Clamps come in a variety of sizes. C-clamps (*Figure 7*), also known as screw jacking, are commonly used because their shape allows them to reach easily around cold-formed steel members. The 9DR locking C-clamp was designed specifically for cold-formed steel framing. Bar clamps are also used extensively to hold steel wall members together until permanent fastenings can be applied.

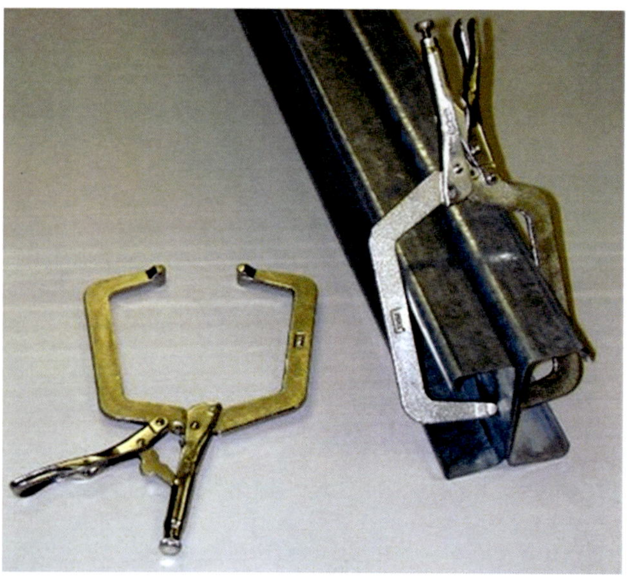

202F07.EPS

Figure 7 ♦ Locking C-clamps.

2.1.0 Cutting Tools

Because so many components of steel framing are prefabricated, the amount of steel cutting that has to be done in the field is minimal. However, when field cutting must be performed, consider using the following tools and methods:

WARNING!

Cutting steel using any of the following methods can result in flying metal filings as well as sharp burrs and edges. Always wear goggles and gloves when cutting steel.

NOTE

Some contract documents prohibit the use of certain cutting tools. Always check installation specifications before using any one method.

- *Chop saw* – Chop saws that have abrasive blades are commonly used for field cutting (*Figure 8*). Abrasive blades grind through metal instead of cutting it. While chop saws are very effective for square cuts and for cutting bundled studs, they are very noisy and produce a rough cut edge. There are a number of circular saws on the market that can be used with carbide-tipped blades to cut steel studs. Carbide-tipped blades provide a cleaner cut through steel than abrasive blades. Some saws have special guards to catch metal chips. This reduces flying debris.

WARNING!

Chop saws give off hot, flying metal filings that can injure workers and create a fire hazard. In addition, the edge produced by a chop saw is very rough, and sharp burrs are left on the steel. Wear gloves and goggles and exercise extreme caution when using these saws.

- *Dry-cut metal cutting saw* – Every year, a number of new or improved tools become available. Dry-cut metal cutting saws (*Figure 9*) fall into this category. These saws use blades made of carbide, titanium, or aluminum that cut through cold-formed steel and leave a smooth edge. The cutting action of this saw is usually slower than that of a saw with an abrasive blade, but the clean cut and lack of hot, flying debris make it safer to use. One manufacturer produces a dry-cut saw that collects the metal filings inside the tool.

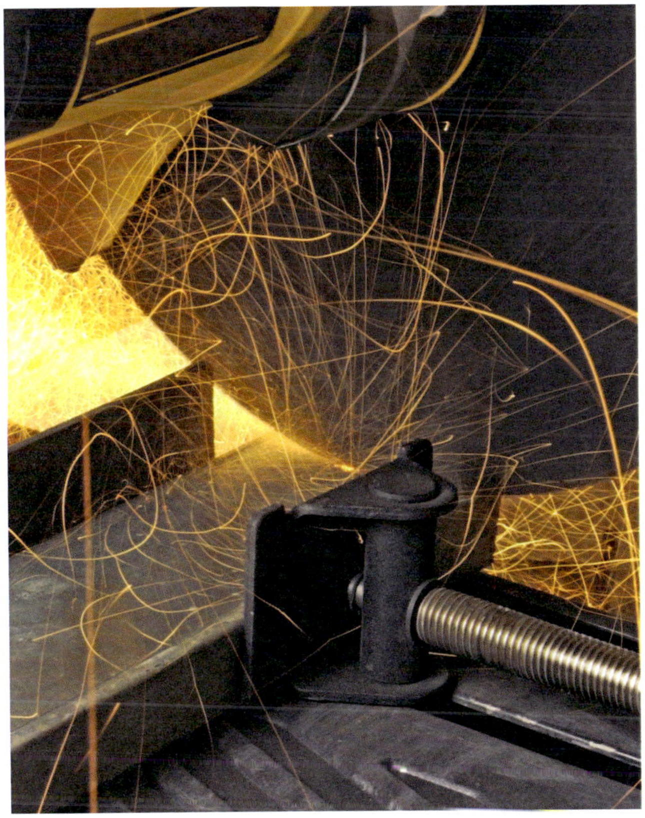

202F08.EPS

Figure 8 ◆ Cutting cold-formed steel with a chop saw.

202F09.EPS

Figure 9 ◆ Dry-cut metal cutting saw.

- *Swivel-head shears* – Swivel-head shears are available in both electric and battery-operated models that can cut many materials up to 68 **mils** thick (33 mils of steel). One mil is equal to $\frac{1}{1,000}$ inch. *Figure 10* shows typical swivel head shears used to cut steel studs, runners, sheet metal, and cold-formed steel. They are portable and make smooth cuts with no abrasive edges. The cutting edge of the swivel-head shear rotates 180 degrees to make overhead and side work easier. Disadvantages include difficulty in cutting tight radii on **C-shapes**, and expensive replacement blades. C-shapes are a cold-formed steel shape used for structural and nonstructural framing members. C-shapes consist of a web, two flanges, and two **lips** (edge stiffeners).
- *Aviation snips* – Aviation snips (*Figure 11*) are hand tools that can cut cold-formed steel up to 43 mils in thickness. They are useful for cutting and coping steel, snipping flanges, and for making small cuts. Some brands of snips are color-coded for left, right, and straight cuts.

CAUTION

Always wear gloves and safety goggles when using snips, because the cut edges are very sharp.

- *Tools for cutting holes* – Most openings in steel studs are prepunched at the factory. However, it may be necessary to punch additional openings for electrical or telecommunications cabling that will be installed later. Since holes in the webs of studs, joists, and tracks affect the strength of a member, all holes must be cut to conform to an approved design or a recognized design standard.

202F10.EPS

Figure 10 ◆ Swivel-head shears.

202F11.EPS

Figure 11 ◆ Aviation snips.

WARNING!

Field-cut holes must follow the requirements in the *Standard for Cold-Formed Steel – General Provisions*. If holes are not cut properly, the integrity of the structural steel members may be compromised.

2.2.0 Laser Devices

Many types of laser devices are used in building construction, including the laser level and laser bob. Both help to make leveling and plumbing tasks easier and more accurate than the manual methods. As with the manual methods, accuracy depends on the skill of the operator. Laser devices emit a high-intensity beam of light that may be visible to the human eye or detected with an electronic sensing device.

The laser level is replacing the conventional level. Fixed beam and rotating beam lasers are available. A rotating beam laser level lets one person do the work that would otherwise take two or more people using a conventional level.

The laser bob (*Figure 12*) is replacing the conventional plumb bob. The laser bob is used to transfer points vertically (from floor to ceiling) with a high degree of accuracy. Most models are self-leveling, increasing their accuracy.

CAUTION

Always exercise caution when using any laser device. Never use a laser device that you are not trained to use. Get instruction first. Some devices require a special operator's training and certification before use. Never shine the laser in anyone's eyes or toward a roadway. Even in daylight, a laser can temporarily blind a person. Always post warning signs in conspicuous locations when a laser is in use in the work area. Although most laser devices do not require the use of safety goggles, always check the manufacturer's recommendations before use.

3.0.0 ◆ FASTENERS FOR STEEL FRAMING WORK

Knowing how to select the right fastener for any steel-framing project will make all the difference in the structural integrity and quality of your work. You must consider the loads to be transmit-

202F12.EPS

Figure 12 ◆ Laser bob.

ted through the connection, as well as the thickness, strength, and configuration of the materials to be joined. Fasteners include screws, nails, pins, clinches, and welds, as well as anchor bolts, rivets, powder-actuated fasteners, and expansion bolts. For applications needing additional stability or when working with steel thicker than 43 mils, the components may be joined by shielded metal arc welding to provide additional strength. It is important to follow manufacturer recommendations for fastening systems during all phases of cold-formed steel construction projects.

3.1.0 Screws

In steel framing, the screw is the workhorse of the fasteners. Screws are installed with screwguns, and they come in a number of different head styles to fit the wide range of structural and cosmetic requirements. Screws have three distinct thread thicknesses: coarse (threads spaced farthest apart), medium, and fine (threads spaced closest together). In steel framing, screws with coarse threads are generally used because they are best for cutting through steel. Thicker steel (for example, 97 mils) requires the use of screws with a fine thread.

Screws are usually finished with zinc or cadmium plating to withstand environmental impacts and resist corrosion. The size and the type of screw needed will depend on the thickness of the sheathing material and steel. Familiarity with the different types of screws that are available is important. If the wrong screw is used, it may not penetrate the steel, or it may break later, causing the joint to fail.

Steel framing is not predrilled for attachment screws. As a result, the screws used to secure steel framing must be able to make their own holes as they are placed. There are two main types of steel framing screws: self-drilling and self-piercing (*Figure 13*). Self-drilling screws do not have threads running the entire length of their shafts to the tip. Instead, they have a drill bit on the tip. Self-drilling screws are designed to cut through layers of steel and make a hole before any of the screw threads engage. Self-piercing screws have threads running the length of their shafts to the tip. The tip is a sharp point that can typically penetrate 18 to 30 mils of material with ease. Self-piercing screws are commonly used to attach plywood and gypsum board to the thinner layers of steel.

Code and standard requirements state that for all connections, screws must extend through the steel a minimum of three exposed threads. For most steel-to-steel connections, ½- or ¾-inch screws are acceptable. When applying plywood, gypsum board, or foam insulation to steel members, the proper screw length is found by adding the measured thickness of all materials and then adding ⅜ inch to allow for the three exposed threads.

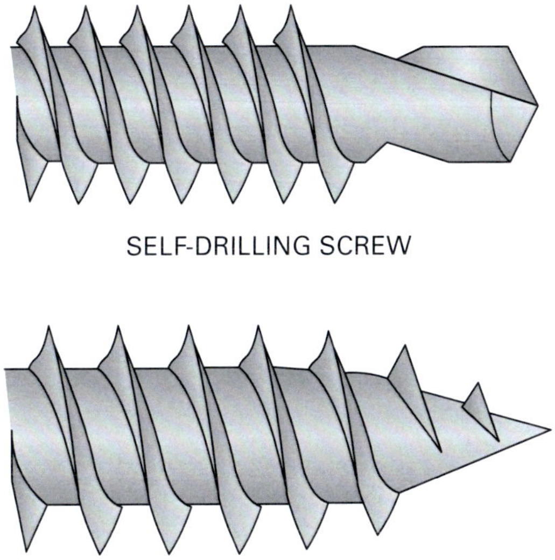

Figure 13 ◆ Self-drilling and self-piercing screws.

CAUTION

When securing screws, over-drilling should be avoided, as it will reduce the effectiveness of the connection and hurt the structural integrity of the sheathing.

Screws are available in many different head styles. The head is used to lock the screw into place and prevent it from sinking through the layers of material it is securing. The head style determines the type of bit tip used to place the screw. The proper selection of a head type profile ensures that the screw will work well with, rather than interfere with, other building components.

Common self-drilling screws (*Figure 14*) include the following:

- *Gypsum board (bugle-head) screw* – Used to attach gypsum board to steel framing. The gypsum board screw should be used with a depth-setting nosepiece on the drill to avoid tearing the gypsum board or protective paper.
- *Flat-head screw* – Used for wood flooring and facings. The flat-head screw countersinks into the wood so that the top of the head is seated flush with the wood surface without causing the wood to splinter or split.

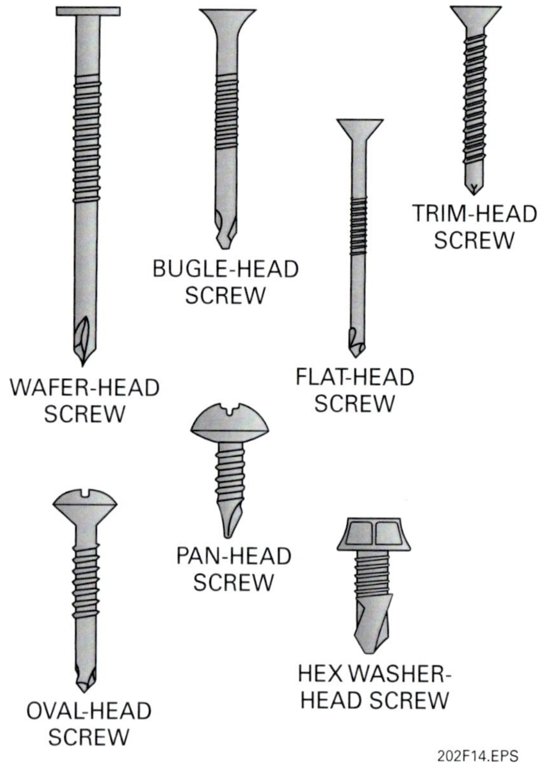

Figure 14 ♦ Typical self-drilling screws.

- *Wafer-head screw* – Has a larger head than the flat-head screw, but it is much thinner. The wafer-head screw seats almost flush with the surface in which it is placed, giving a clean, finished look. The large head gives it a greater surface area than other screws. For this reason, it is used for attaching to steel studs those materials that can be easily torn, such as felt or reflective foil.
- *Hex washer-head screw* – The head is hex-shaped, and is also flared at the bottom, which serves as a washer. These screws are placed using a hex-head driver socket. The driver socket fits securely over the screw head, providing good stability while the screw is being driven into place. If sheathing and gypsum board are placed over the fastener, the head style must have a very thin profile to prevent blow-outs or bumps at the screw locations.
- *Trim-head screw* – Small, trim-head screws are used when baseboards and trim are attached to steel studs. The small head penetrates the trim, leaving a tiny hole that is easily filled with putty. These screws may be used with pneumatic and powder-actuated trim nailers.
- *Oval-head screw* – Has a head that is shaped so it entirely covers a clearance hole. In addition, the head has a low profile and is inconspicuous. The oval-head screw is used to attach fixtures, such as shelves and cabinets, to steel stud walls.
- *Pan-head screw* – Used to fasten studs to runner tracks. The pan-head screw is also used to connect steel bridging, strapping, or furring channels to studs or joists.

3.2.0 Pins

Pins resemble nails that have a spiraled or knurled shank, as shown in the first example of *Figure 15*. Pins can be used for steel-to-steel connections or to attach sheathing material to steel members. Connections are made by driving the pin into the layers of material. Nail guns are commonly used to attach plywood, oriented strand board (OSB), and siding to steel framing. Some contractors use only

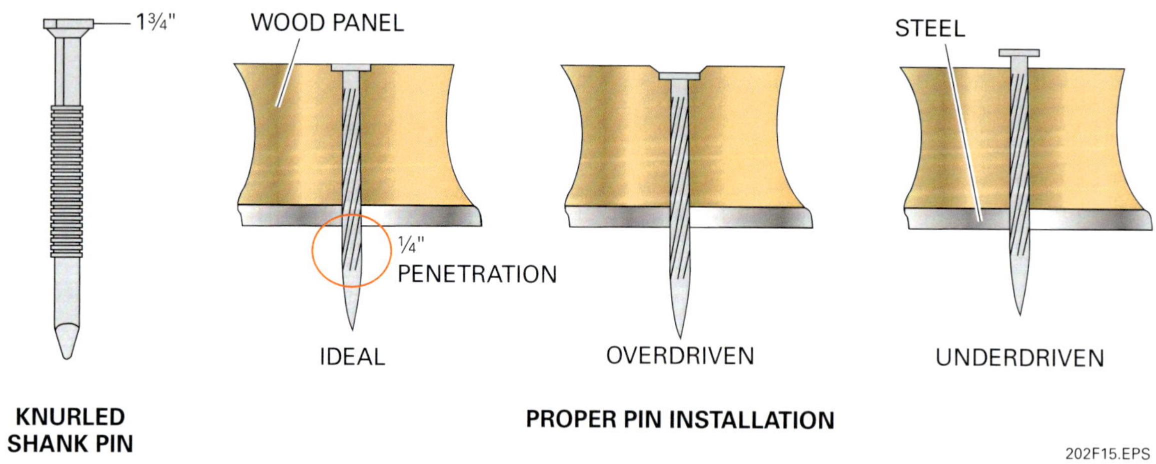

Figure 15 ♦ Pin types and pin installation.

nails (or pins), while others use screws around the edges and pins in the field of the board. Plywood or OSB sheathing material may be applied using pneumatic pins, as shown in the second example of *Figure 15*. To securely attach the sheathing, it must be held tightly against the steel before the pin is driven. This is because firing the pin does not pull the plywood against the steel as a screw does. Roof sheathing is more easily installed with pins than wall sheathing because, on the roof, the worker usually stands on the plywood to hold it tightly against the steel. Sheathing may also be attached to steel studs by first tacking the material to the steel with screws along the edges and then applying pins to the field.

3.3.0 Clinching

Clinching requires no screws or pins, only a pneumatic tool that press-joins pieces of steel together. The clinching tool, in effect, creates a rivet in the metal, as shown in *Figure 16*. Manufacturers of clinching tools provide test reports that verify the strength of the press-joining. Clinching systems work well on steel panels and are less permanent than welding. If studs are fastened incorrectly with a clinch connection, the connection may be popped out with a screwdriver, or it may be drilled out.

3.4.0 Bolts and Anchors

Bolts and anchors are also commonly used to fasten cold-formed steel framing to masonry, concrete, and other steel components. Except for some proprietary anchors, pre-drilling of holes is necessary. Bolts require the installation of a washer, and must meet or exceed the requirements of *ASTM A307*. Expansion anchors are commonly used for connections to concrete or masonry and require information from the manufacturer to determine the capacity and spacing requirements.

4.0.0 ♦ STEEL FRAMING MATERIALS

Framing components include steel studs, steel joists, and steel roof trusses. Framing systems come with accessories such as clips, web stiffeners, resilient channels, fastening devices, and anchors required for complete and proper installation of members. Many manufacturers offer specialized products that enable builders to shorten construction times for complicated curves, arches, and other unique architectural features. The vertical and horizontal framing members serve as structural load-carrying components for a large number of low- and high-rise structures.

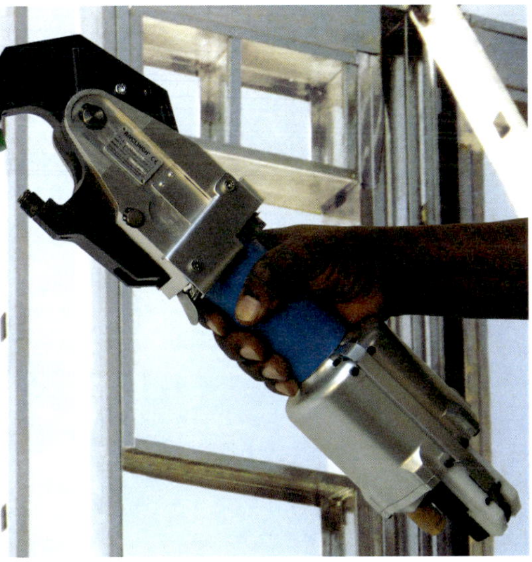

CLINCHING TOOL

CROSS-SECTION OF CLINCHED STEEL

202F16.EPS

Figure 16 ♦ Pneumatic clinching tool.

Steel framing materials have several advantages over wood-framing, such as lower costs, faster assembly and installation, and design flexibility. The components are also non-combustible, lightweight, corrosion-resistant, and fabricated to fit together easily. Manufacturing tolerances for cold-formed steel members are governed by ASTM standards. These standards govern length, web depth, flare, crown, bow, and twist. Cold-formed steel components are also coated against corrosion according to ASTM standards. Hot-dipped zinc galvanizing is the most effective coating method. Depending on the thickness of zinc applied to the steel and the environment in which the steel is placed, zinc coatings can protect the steel for more than 1,000 years.

There are a variety of loadbearing and non-bearing systems. *Figure 17* shows the basic components of a steel stud system. To meet custom material requirements, studs, track, and joist material can be cut within ⅛ inch of specifications. Length is restricted only by the mode of physical transportation—typically 40 feet for containers and flat bed trucks. Custom ordering allows for less field cutting and less labor and waste on the job site.

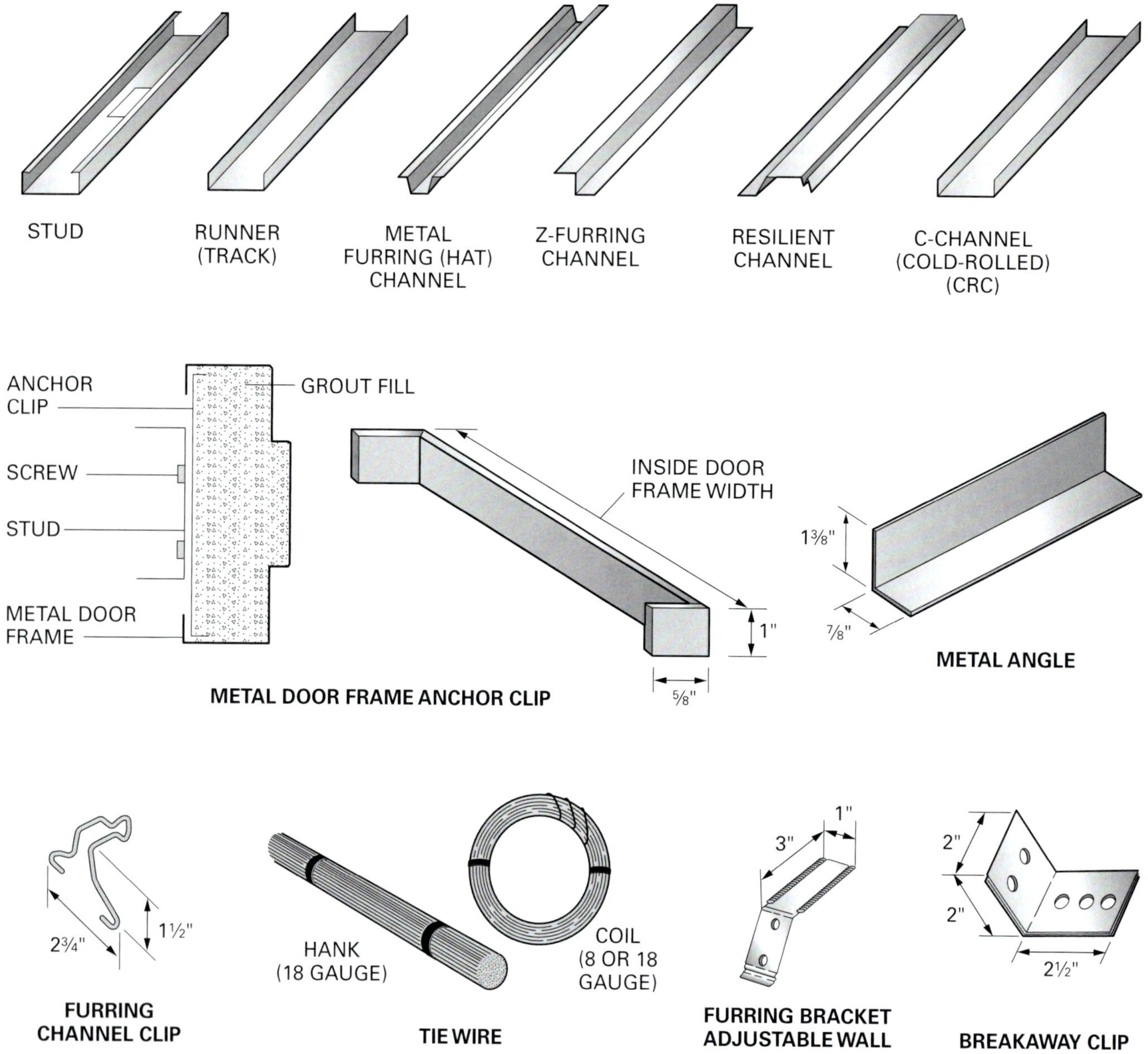

Figure 17 ◆ Components of a metal (steel) stud system.

Steel framing members that are not marked with the required identification should not be used. The identification may be etched, stamped, or labeled every 96 inches along the center line of the web (*Figure 18*). Missing identification can result in the project delays until materials are identified in accordance with appropriate standards.

4.1.0 Identification of Framing Materials

Manufacturers of various steel framing products have published widely varying codes for identifying steel components. To eliminate this confusion,

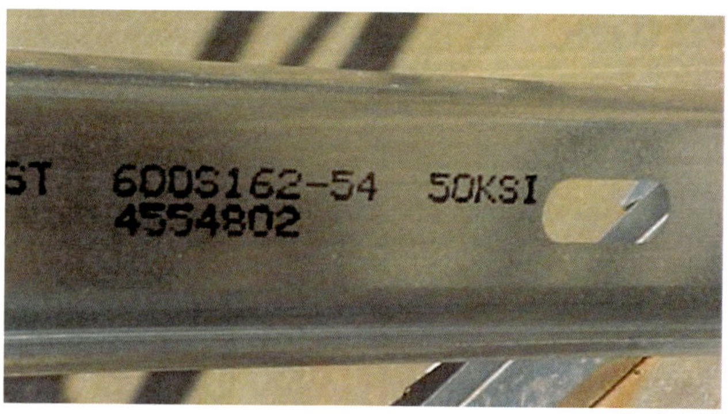

Figure 18 ◆ Steel product label.

GOING GREEN

Recycled Steel

Modern steel production relies on two technologies: the basic oxygen furnace (BOF) or the electric arc furnace (EAF). The BOF process uses 25 to 35 percent old steel to make new steel. It produces sheet for products in which the main requirement is drawability; that is, products like automotive fenders, steel-framing members, refrigerator encasements, and packaging. The EAF process uses 95 to 100 percent old steel to make new steel. It is primarily used to manufacture products in which strength is critical, such as structural beams, steel plates, and reinforcement bars. No matter which process is used, the resulting steel product has a minimum of 25 percent recycled content. Industry-wide, the recycled content of steel averages 67 percent.

During the production process, molten steel is poured into an ingot mold or a continuous caster, where it solidifies into large rectangular shapes known as slabs. The slabs are then passed through a machine with a series of rolls that reduce the steel into thin sheets at desired thickness and strength, and having other physical properties. The sheets are sent through a hot-dipped galvanizing process, and are then rolled into coils that weigh approximately 13 tons.

the Steel Stud Manufacturers Association (SSMA), representing more than 80 percent of all U.S. steel framing production, developed a universal designator system. The STUFL system has become the industry standard for identifying the most commonly used framing members. The designator consists of four sequential codes. The first is a three- or four-digit number indicating the member web depth (D) in 1/100 inch. The second is a single letter indicating the type of member:

- S = Stud or joist framing member with lips
- T = Track section
- U = Channel or stud framing section which do not have lips (also known as edge stiffeners)
- F = Furring channels
- L = Angle or L-header

Using *Figure 18* as an example, the designator system for a 6-inch, 16-gauge C-shape with 1⅝ inch flanges is 600S162-54:

- 600 = 6-inch member depth expressed in 1/100 inch (outside to outside dimension)
- S = Stud or joist with flange stiffeners
- 162 = 1⅝-inch flange in 1/100 inch
- 54 = Minimum base steel thickness in mils (0.054 inch = 54 mils)

Base steel thickness was traditionally referenced in terms of gauge numbers, but that has given way to designating thickness in terms of mils. One mil is equal to 1/1,000 of an inch. *Table 1* lists common base steel thicknesses in mils and equivalent gauge identifiers. More information about STUFL can also be found in *Appendix B* at the end of this module.

Table 1 Minimum Base Steel Thickness of Cold-Formed Steel Members

Designation (thickness in mils)	Minimum Base Steel Thickness Inches (mm)[1]	Old Reference Gauge Number[2]
18	0.0179 (0.455)	25
27	0.0269 (0.683)	22
30	0.0296 (0.752)	20 – Drywall[3]
33	0.0329 (0.836)	20 – Structural[3]
43	0.0428 (1.09)	18
54	0.0538 (1.37)	16
68	0.0677 (1.72)	14
97	0.0966 (2.45)	12
118	0.1180 (3.00)	10

[1] Design thickness shall be the minimum base steel thickness divided by 0.95.

[2] Gauge thickness is an obsolete method of specifying sheet and strip thickness. Gauge numbers are only a rough approximation of steel thickness and shall not be used to order, design, or specify any sheet or strip steel product.

[3] Historically, 20 gauge material has been furnished in two different thicknesses for structural and drywall (non-structural) applications.

202T01.EPS

4.2.0 Basic Construction Methods for Steel Materials

There are requirements and guidelines that apply to cold-formed steel construction for all elements within a building. These elements include in-line framing, web stiffeners, standard web holes and patches, and stud-to-track gap tolerances. Steel frame assemblies also require certain types of bracing for stability.

4.2.1 In-Line Framing

In-line framing, or direct alignment, is the preferred and most common framing method for providing a direct load path for transfer from studs to joists, through the framing system, and to the ground. In this method, cold-formed steel framing members are aligned vertically so that the center line of the joist web is within ¾ inch of the center line of the structural stud member below, or the center line of the stud web is within ¾ inch of the center line of the web joist below (*Figure 19*).

4.2.2 Web Stiffeners

Web stiffeners, or bearing stiffeners (*Figure 20*), in the shape of C-shaped members or track members, are used to prevent joists from crippling at the point where the load transfers from a stud into the floor joist under structural walls. The thickness of a stiffener is, at a minimum, the same thickness as the floor joist, and the length of the stiffener is the depth of the joist minus ⅜ inch. Stiffeners are installed across the joist depth of the web and on either side of the web. The stiffener is fastened to the web with either three or four No. 8 screws. Three fasteners are used when the screws are installed in a single row, and four are used when screws are installed with one fastener in each corner.

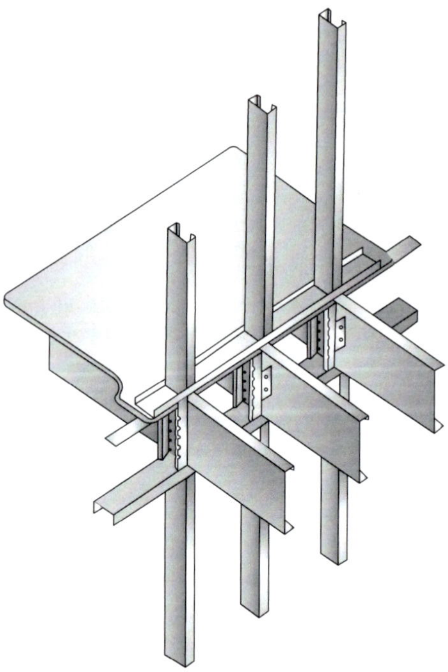

202F19.EPS

Figure 19 ◆ In-line framing detail.

202F20.EPS

Figure 20 ◆ Web stiffener.

4.2.3 Web Holes and Patches

Depending upon the application, framing members may be designed and manufactured with standard punchouts—web holes made during manufacturing (*Figure 21*). Holes in webs of studs, joists, and tracks must conform to an approved design. Standard web hole sizes are typically 1½ inches by 4 inches, and are located on the center line of the web of the member at 24 inches. Propri-

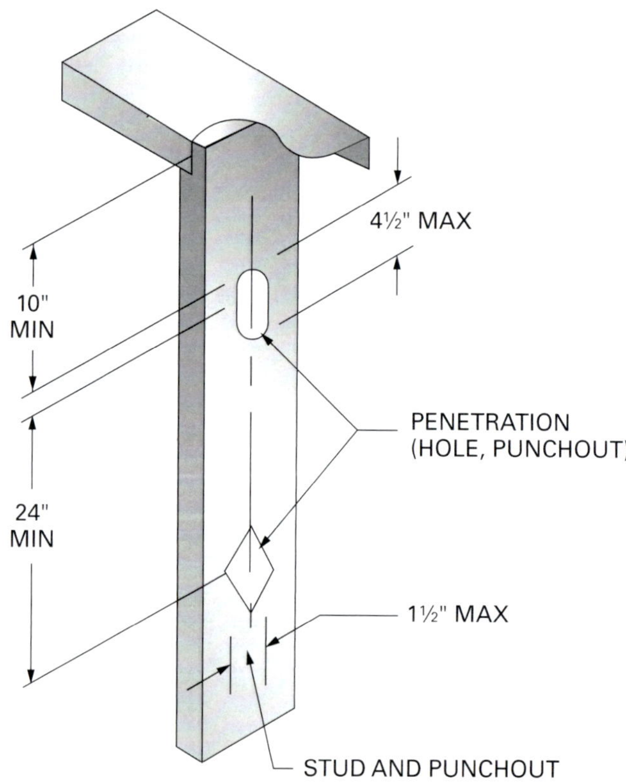

202F21.EPS

Figure 21 ◆ Standard stud punchouts.

etary products are also produced with much larger holes that have been engineered and approved to ensure that the member's strength is not compromised.

CAUTION

Do not patch or reinforce hole errors yourself. If a structural member is damaged, consult a design professional for corrective action.

WARNING!

Field modifications to framing members cannot be attempted without the approval of the design professional. Damaged materials should never be installed, because they may compromise structural integrity.

4.2.4 Built-Up Shapes

The open, C-shaped cross-section commonly used in cold-formed steel framing provides a very low resistance to twisting. Built-up shapes, such as a nested stud and track assembly, provide the increased stiffness of a closed section as well as flat surfaces for attaching finish materials. Built-up shapes are commonly used for door and window jambs, headers, beams, and posts (*Figure 22*). The strength of built-up shapes is determined by the properties of the individual members, as well as the method of fastening.

4.2.5 Bridging, Bracing, and Blocking

The strength of individual framing members is a function of the bracing provided. This bracing restrains the member from moving laterally or twisting, and it may be in the form of any one or combination of the following:

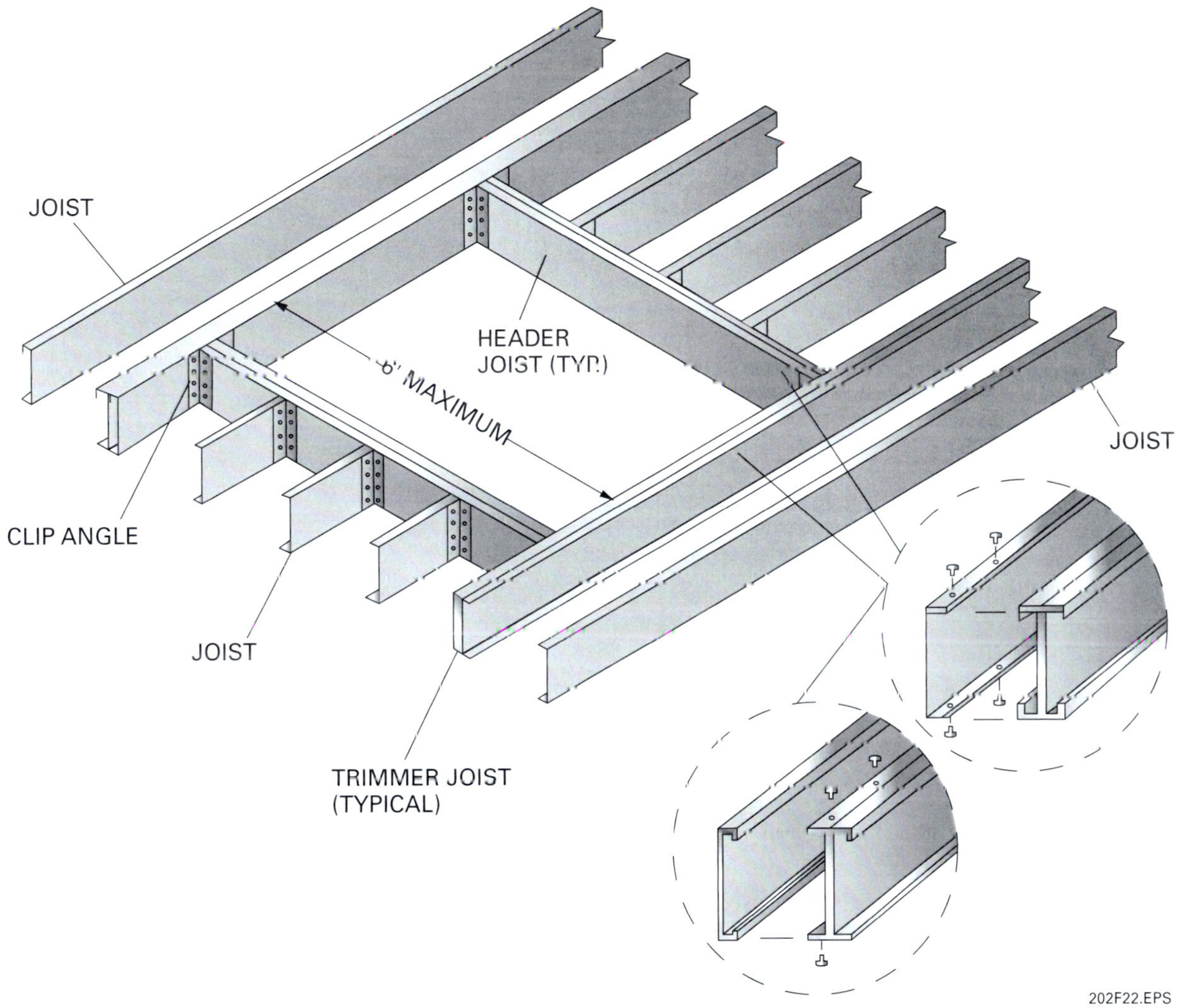

Figure 22 ♦ Floor opening.

- Cold-rolled channels placed through the web punchouts
- Steel strapping attached to the flanges with periodic solid **blocking**, X-bracing
- Wood sheathing attached to the flanges

Overall, system bracing anchors the steel members and provides stability to the entire structure.

4.2.6 Thermal Considerations

Products used in construction may require some additional insulation to meet energy codes. *The Thermal Design Guide for Exterior Walls*, produced by the American Iron and Steel Institute, provides designers and contractors with guidance on thermal design of buildings that use cold-formed steel framing members. The thermal performance of a steel-framed structure may also be improved by the batt and other insulating materials within the wall cavity. Some thermal regions require insulation foam board on the exterior of the frame. Designers should consider the effects of moisture when assessing the application of cavity and continuous insulation.

4.2.7 Pressure-Treated Wood

Be careful when using pressure-treated wood products with cold-formed steel framing, as accelerated corrosion may result. It is preferable not to use pressure-treated wood with steel framing, but if you do, specify a less corrosive pressure treatment, such as sodium borate. Always separate the steel framing material from the pressure-treated wood with a nonabsorbing, closed-cell sill seal.

4.2.8 Custom Jigs

The use of identical components, hardware, and accessories allows buildings to be constructed quickly and with little waste. For example, most doors in a building are identical. This saves time during construction. Each door opening will be the same size, each door jamb will be constructed in the same way, and all doors will use the same types of hardware.

When the same task is performed over and over, it is often helpful to construct a jig to help do the job. Examples of jigs include the following: a wooden frame that holds a steel stud upright while a permanent connection is being made; a template that is used to accurately drill holes in a steel member; and a frame that is used to square connections as the frame is being assembled. Jigs make doing the job faster and easier.

4.3.0 Assembling a Steel Frame

The following section is designed to provide you with a basic understanding of how to assemble steel framing. If you are familiar with wood framing, you will notice that the fundamental concepts of structural design are almost identical with those of steel framing. Detailed information about assembling steel framing is provided in other sections of this module.

Appendix C contains a number of diagrams showing details of various types of steel framing. Keep in mind that when working with steel framing, you will likely be coordinating tasks with other crafts. For example, electrical wiring must be placed before wall sheathing.

4.3.1 Stick Framing

Many builders choose to frame walls in place, especially when the floor is not level enough to build panels. To frame walls in place, cut the track for the full length of the wall. Mark the top and bottom track for layout, and anchor the bottom track in place, securing the studs in each end of the track. Position the top track at the ends and with intermediate studs. Use a string line and level to position the remaining studs for the wall. Install headers, X-bracing, or plywood with the wall standing.

Figure 23 provides a basic wall framing design. This type of wall may either be built in place or built horizontally and then raised into position and secured to the building foundation. The spacing between the studs depends on the design of the building, the loads that will be placed on the structure, and the gauge of the steel being used. Typical stud spacing is 16 or 24 inches OC. Studs are bridged or strapped whenever there is a possibility that they will twist.

Depending on the design of the wall, it may need additional stabilizing. **Shear walls**, which need to withstand forces from wind and earthquakes, may need diagonal bracing. Other walls, such as interior nonbearing walls, may not require additional bracing; the sheathing attached to the studs may provide enough stability.

The possibility of vertical and horizontal deflection often figures into decisions about how the upper track will be secured to the wall studs. Sometimes the studs are screwed to the upper track, but this method does not allow for movement. The screw can become a pivot point when stress is applied to the wall, causing the joint to fail. More often, a method that allows for movement is used. One technique is to use slotted tracks (*Figure 24*). The screw holes in the sides of

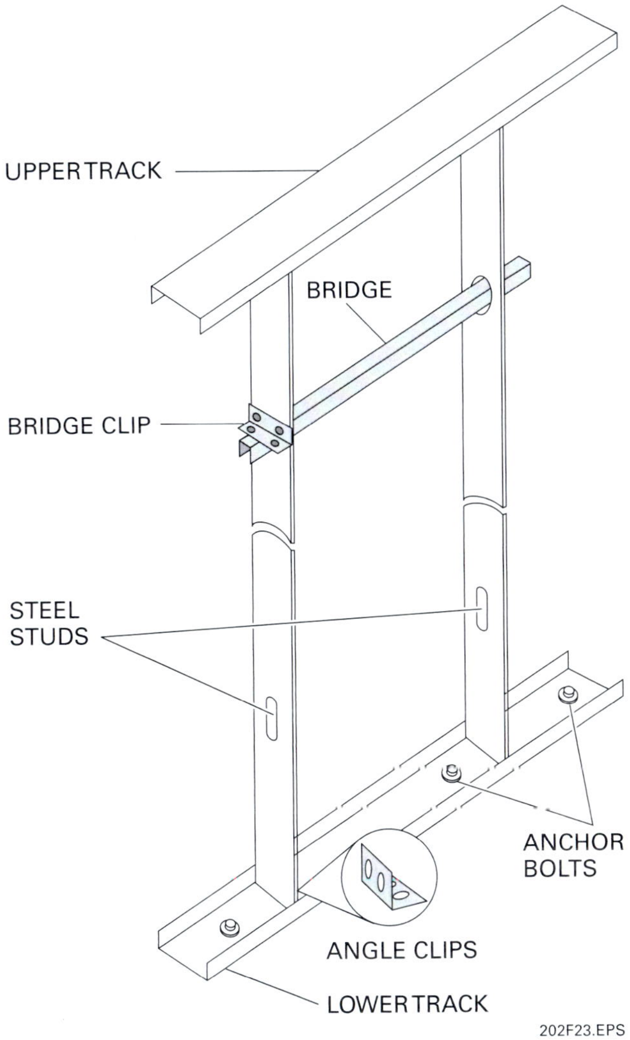

Figure 23 ♦ Basic steel wall assembly.

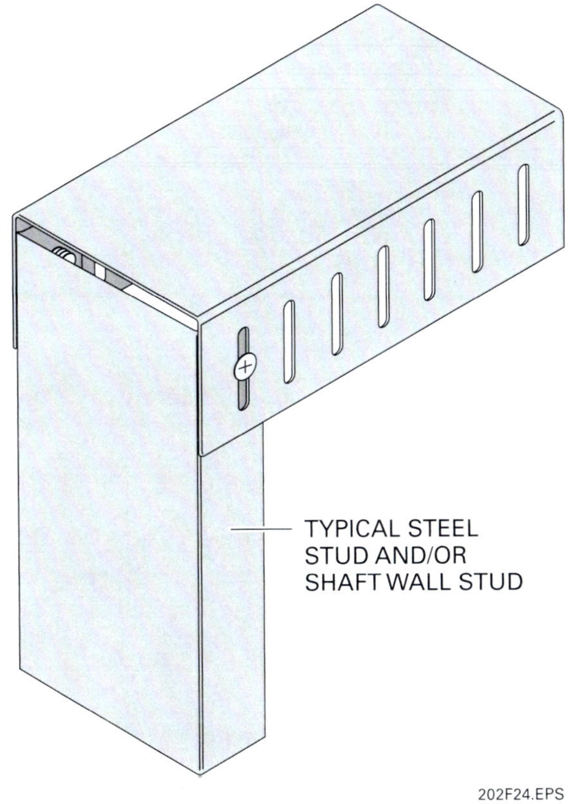

Figure 24 ♦ Slotted track.

the track are vertical slots. The stud is attached to the track using a connector with some type of bushing that allows free movement of the connector in the track. When the stud is attached to the track, it can move up and down within the slot as the floor or roof above is placed under load.

Another method is to use deep leg tracks. In this design, the sides of the track are deeper than those of a standard track. The deep leg tracks are placed over the tops of the studs, but the studs are not secured to the upper track. As a result, when the upper part of the structure is under load, the track moves down until it rests on the tops of the studs. When this method is used, the studs must be properly stabilized to prevent twisting and bending.

A third technique is called slip track. In this method, the tops of the studs are attached to the upper track. The upper track is then nested in a second track, with no connection between the two tracks. The inner track is attached to the studs, which helps to stabilize the wall. The inner track floats inside the outer track as a vertical load is applied.

Another process uses a variety of clips to connect members together (*Figure 25*). The clips are formed into a 90-degree angle, and slotted connector holes allow the connectors to move freely as a load is applied to the member. The different designs of the clips allow for horizontal (drift) or vertical deflection, or both. The clips may be used in a track, or they may be attached directly to stud and floor members.

Corner clips are another option that designers sometimes use to allow for side-to-side movement of building corners (*Figure 26*).

4.3.2 Panel Framing

Studs are placed in the lower track of the frame. The studs must be seated tightly into the track so the bottoms of the studs rest on the bottom of the track. The studs are then attached to the track with a specified angle clip or with screws that are driven through the side of the track and the flange of the stud. Stud bottoms need to rest on the floor of the track so that the weight placed on the stud is transferred to the track rather than being placed on the screw connection.

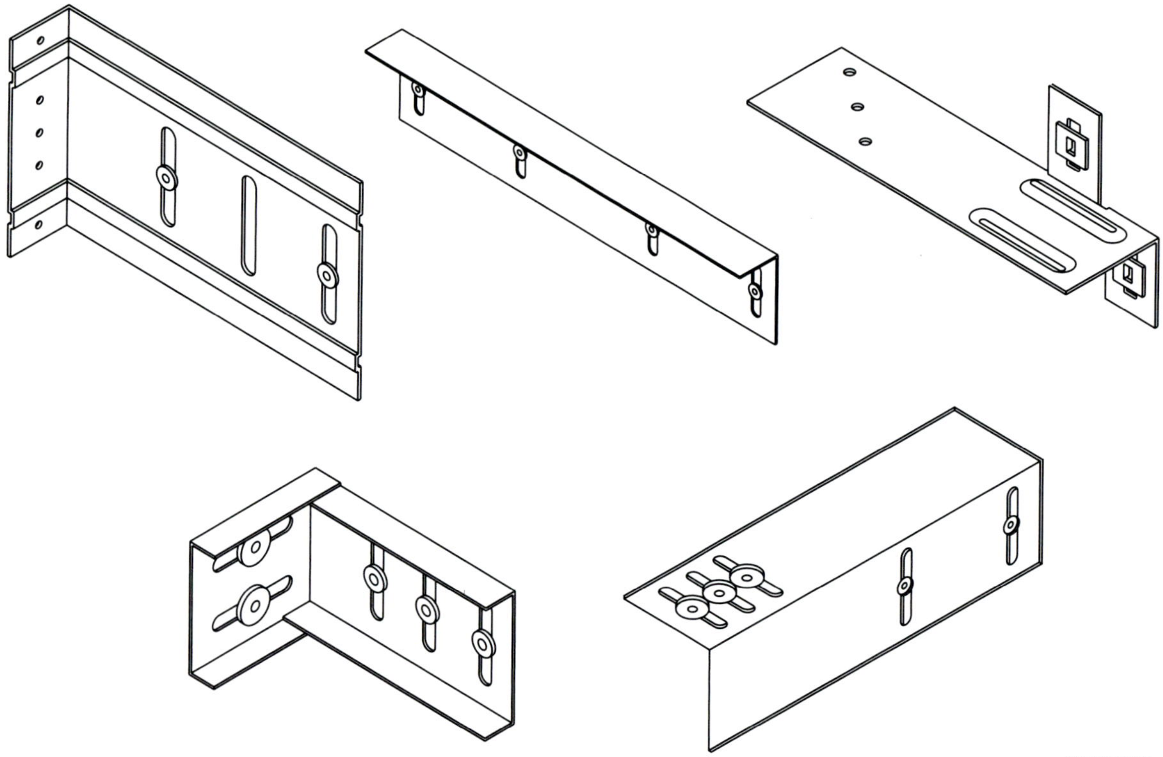

Figure 25 ◆ Clips that allow for drift.

The tops of the studs are capped with another track. Just as they are in the bottom track, the studs need to be placed snugly against the top track. This ensures that any weight applied to the track is transferred to the studs, not to the connection. The upper track may also be used for attaching floor joists for the next floor, ceiling joists for the roof trusses, or roof members.

When the frame is raised and placed, it is attached to the building foundation by means of the lower track. Anchor bolts (*Figure 27*) are frequently set into the foundation before pouring the concrete. They may also be drilled in after the concrete has cured. Anchor bolts allow the lower track of the steel framing to be bolted or welded to the foundation. Powder-actuated fasteners may also be used to attach the lower track to the foundation.

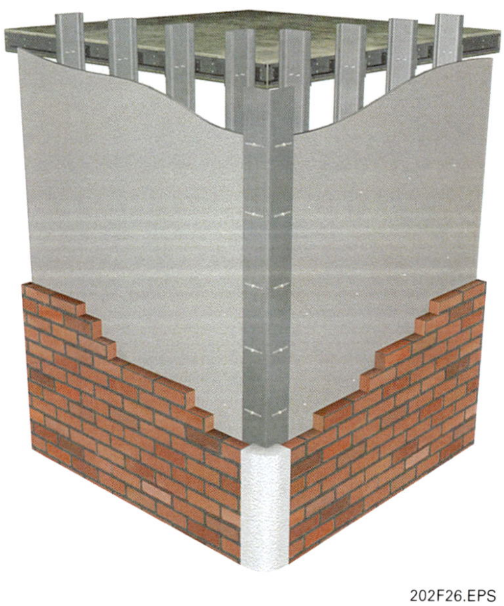

Figure 26 ◆ Corner clip.

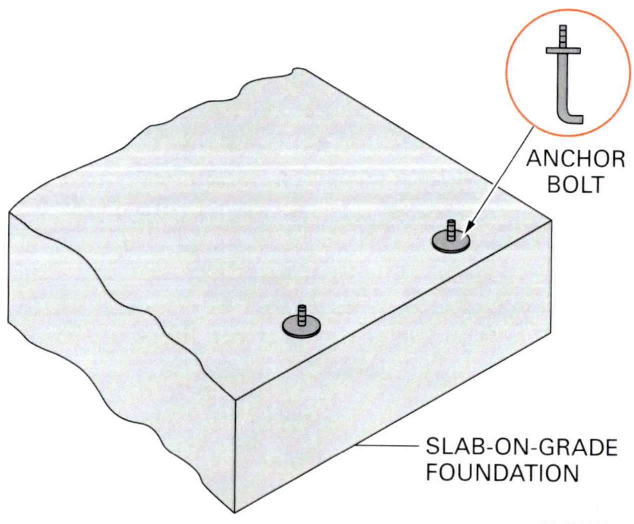

Figure 27 ◆ Typical anchor bolt installation.

5.0.0 ♦ STEEL FRAMING APPLICATIONS

Historically, cold-formed steel framing has been widely accepted for use in nonstructural applications or partition walls. Today, it is commonly used for floor, structural wall, and roof assemblies as well. This section covers steel framing as it applies to the structural walls used in residential and commercial structures (*Figure 28*).

5.1.0 Framing Walls

Structural walls support the weight of the building and protect occupants from wind loads and other forces of nature. A basic, cold-formed structural steel wall includes structural studs, tracks, fasteners, bracing, and bridging. Headers for window and door openings may be constructed using structural studs, joists, or proprietary cold-formed shapes. In addition, shear walls are typically framed within the stud wall assembly.

Wall framing members have a material base steel thickness of 33 to 118 mils, with a minimum metallic zinc coating of G60. Typically, structural wall studs are produced in sizes from 2½ to 8 inches in width, with flanges ranging from 1⅝ to 2½ inches or more. The return lips depend on the flange size. Structural track is sized to accommodate the web depth of wall studs, and the flanges are typically sized 1¼ to 3 inches. The selection of a particular member will depend on its intended use. A number of proprietary and nonproprietary accessories are also available for structural wall assembly, including cold-rolled channel for bridging, as well as clips, angles, and straps.

202F28.EPS

Figure 28 ♦ Examples of structural wall systems.

5.1.1 Layout

It is important to lay out the wall studs accurately so that they align with other structural assemblies such as roof and joist framing. Steel framing is typically spaced at 24 inches on center to take full advantage of the framing member.

Place the top and bottom track members on the straight edge of the panel table. They should be arranged with the webs next to each other, clamped together temporarily. They should fit tightly against the edge of the end stop (the straight edge at the end of the wall). Mark the layout of the wall studs on the flanges of the top and bottom track,

INSIDE TRACK

Prefabricated Panel Building Construction

Today, a number of companies are offering services for prefabricating panelized cold-formed steel exterior and interior structural walls, interior partitions, floors, and roofs in accordance with architectural plans. The prefabrication is accomplished off-site in a factory and the panelized components are shipped assembled to the construction site. Some companies provide prefabrication of the panelized components on-site in mobile facilities. In either case, stick-fabrication of the building is virtually eliminated.

starting with a wall stud at the end of the wall. Use highly visible ink, such as a black felt-tip marker. Place a line at the web location and an "X" on the side of the line to indicate the stud flanges.

Mark the next stud location to match the first truss or roof rafter location from the end wall. Continue marking every 24 inches (or 16 inches, depending on the layout) for the full length of the wall. Where the exterior corner walls intersect, the wall that runs to the edge of the foundation has an extra stud. This stud is 3 inches from the end and acts as a backer to screw to the shorter intersecting wall.

Next, identify the rough openings in the walls. Check the architectural drawings to find door and window locations. If the dimensions are not provided, scale them from the drawing. Mark the location for the center of the openings on the top and bottom tracks. Use a red felt-tip marker to distinguish these marks from the layout marks. Check the door and window sizes on the drawings and verify rough openings with actual window sizes. Add 12 inches to the width of the window openings to allow for two jack studs on each side of the header with room to spare.

Using a tape measure, center the dimensions over the red marks on the track. Mark each end of the tape measure. This shows the location of the webs of the king studs. Put the "X" on the side of the mark away from the window. The webs of the king studs will be on the rough opening side. Remember to mark both the top and bottom track with 12 inches added to the rough window openings. This dimension is standardized to simplify header ordering and assembly. Two king studs may not be required at every opening, and framers may vary the length of the headers. However, standardizing header lengths helps to simplify cut lists.

All loadbearing studs must be aligned with the trusses, joists, or rafters above or below the wall. Because these walls carry loads, it is important to fit each stud tightly into the track member to allow the stud to properly carry the axial (downward) load from above. The top and bottom wall tracks must be of equal or greater thickness than the studs. Panelized walls may be constructed on a concrete slab, floor deck, or panel table. For all systems of wall construction, make sure that the surface on which the wall is built is level. Walls must not be out of plumb more than ⅛ inch for every 10 feet.

Walls may be framed in two ways. Full-length walls may be framed up to 40 feet depending on available manpower at the job site. Be aware that a longer wall also tends to twist, and could get damaged if the crew is not large enough to keep the wall straight. Shorter wall sections need more plumbing and alignment, but they are easier to assemble and may be built and spliced together. The track will also need to be spliced (*Figure 29*).

5.1.2 Wall Assembly

Before beginning assembly of a structural wall, make sure that the foundation or the bearing surface is free of all defects. The bearing surface should be uniform, with a maximum ⅛-inch gap between it and the track.

After layout has been completed, separate the top and bottom tracks, and install a wall stud at each end of the wall between the top and bottom tracks. Clamp the stud flanges to the track flanges with locking C-clamps at each end. Tap the track on one end with a hammer to seat the studs as tightly as possible in the track. Fitting each stud tightly and perpendicular to the track keeps the wall straight. Screw one low-profile No. 8 screw through the flange of the track into the flange of the stud on both sides of the stud. If an elevated panel table is used, the framer may be able to install the screw (from underneath) on the other flange as well. If not, once all the studs and headers are in place and all screws are installed on one side, flip the wall over to install the screws on the other side.

In drywall framing, the framer often only installs screws on one side of the wall. Check local codes. However, for loadbearing walls, screws must be installed on each flange on both sides of the wall. This keeps the studs from twisting and provides the proper connection for in-

202F29.EPS

Figure 29 ◆ Track splicing.

line framing. Continue twisting the studs into the track. Install all the studs the same way, with the open side of the C-shape facing in the same direction and toward the start of the layout. Align the punchouts in the studs to provide straight runs for the plumbers and electricians. The studs should be aligned so that they all face in the same direction on parallel loadbearing walls.

Install the king studs at the rough openings with the hard side of the stud facing the rough opening and punchouts aligned. Do not install studs at the markings between the king studs. These markings will be for the cripple studs when the rough openings are framed. Continue down the length of the wall until all the studs are twisted and screwed into place. Do not remove the wall panel from the table until the headers and rough openings have been completed.

During wall assembly, it is important to seat the studs into the track as tightly as possible. Studs must be seated with a gap of no more than ⅛ inch. This ensures that the building loads are transferred through the studs, and not through the fastener (*Figure 30*).

Structural walls may be constructed using any of the following methods:

- *Stick building* – Walls are framed on the job site, one stud at a time.
- **Panelization** – Walls are built on a flat surface at the site, such as a concrete slab.
- *Pre-engineered method* – This assembly typically increases the size and spacing of structural steel members. In some cases, the spacing can be as much as 4 or 6 feet on center.

INSIDE TRACK

Advantages of Metal Framing

Metal framing offers a great deal of flexibility in building design. Fire ratings are easier to achieve because, unlike wood, steel framing is noncombustible. In addition, steel framing can be produced in lengths not readily available in wood.

202SA01.EPS

ACCEPTABLE UNACCEPTABLE

202F30.EPS

Figure 30 ◆ Stud seating.

Screws are commonly used to attach track to cold-formed steel floor or roof assemblies, but welding may be used when specified. Powder-actuated fasteners may be required when the track connects to structural steel-framing members or concrete. Expansion bolts are used at jamb locations and corners, while expansion anchors are typically used at shear wall locations.

5.1.3 Wall Installation

If anchor bolts are used in the foundation, measure their locations and place holes in the bottom track of the wall panels so that the walls will fit over the bolts. If strap anchors or other kinds of anchors are used, this step will not be necessary. Place temporary bracing material near the foundation in preparation for the wall to be raised. Any stud material at the job site, preferably 12 feet long, may be used. Caulk the concrete foundation with weatherproof caulking material and use foam, closed-cell sill sealer.

Move the wall panel and set the bottom track on the foundation. Position it over the anchor bolt locations and tilt the wall up. Leaving the wall tilting slightly outward, clamp the temporary brace material to the wall studs in two or three locations, depending on the length of the wall. Make sure the braces do not lap past the inside face of the wall. Before removing the clamps, secure a No. 10 hex head screw through the brace into the stud. Install a brace every 8 to 12 feet along the wall. Secure the bottom of the brace with a stake driven into the ground or other solid surface. Screw through the stud into the stake to hold it in place. Repeat this process with all wall panels until all loadbearing walls are standing.

WARNING!

Before tilting up a wall assembly, make sure that the braces do not hang past the wall edges. Snagging on the braces is very dangerous and could result in injury.

A balloon frame can be used for some structures (*Figure 31*). This is categorized as a pre-engineered structure. Strictly adhere to the engineered drawings.

5.1.4 Shear Walls

Lateral (shear) loads are typically resisted by a system of interconnected shear walls and floor and roof/ceiling **diaphragms** that work together to transfer applied wind and seismic loads to the foundation of the structure. These diaphragms are called shear walls. Shear walls can be created by attaching gypsum board, wood structural sheathing, steel decking, X-bracing (*Figure 32*) or other materials to the floor, wall, ceiling, or roof framing. In addition to field-fabricated shear wall systems, there are now several proprietary, high-strength pre-engineered systems that may help shorten construction time (*Figure 33*).

202F31.EPS

Figure 31 ◆ Balloon framing.

The wall design for any steel-framed structure should provide for the placement of shear walls. In multi-story construction, proper alignment (stacking and load path) is also required to ensure that there is adequate shear transfer between roof or floor diaphragms and shear walls, or other assemblies such as bar joists, long-span deck, and wood framing. Shear walls are usually connected to the foundation using proprietary or engineered connectors and hold-downs.

NOTE

All concealed cavities such as headers in exterior walls must be pre-insulated before they are installed.

202F32.EPS

Figure 32 ♦ X-bracing.

202F33.EPS

Figure 33 ♦ Proprietary shear wall.

5.1.5 Header Assembly

Loadbearing headers are typically boxed, unpunched C-shapes that are capped on the top and bottom with track sections (*Figure 34*). They must be engineered for bending and shear strength, along with web crippling (crushing) at the locations of the loads from above. The two types of headers that are most commonly built from standard C-shapes are the box and back-to-back headers. A third type, called an L-header, is a steel angle in the shape of an L. Prefabricated headers may also

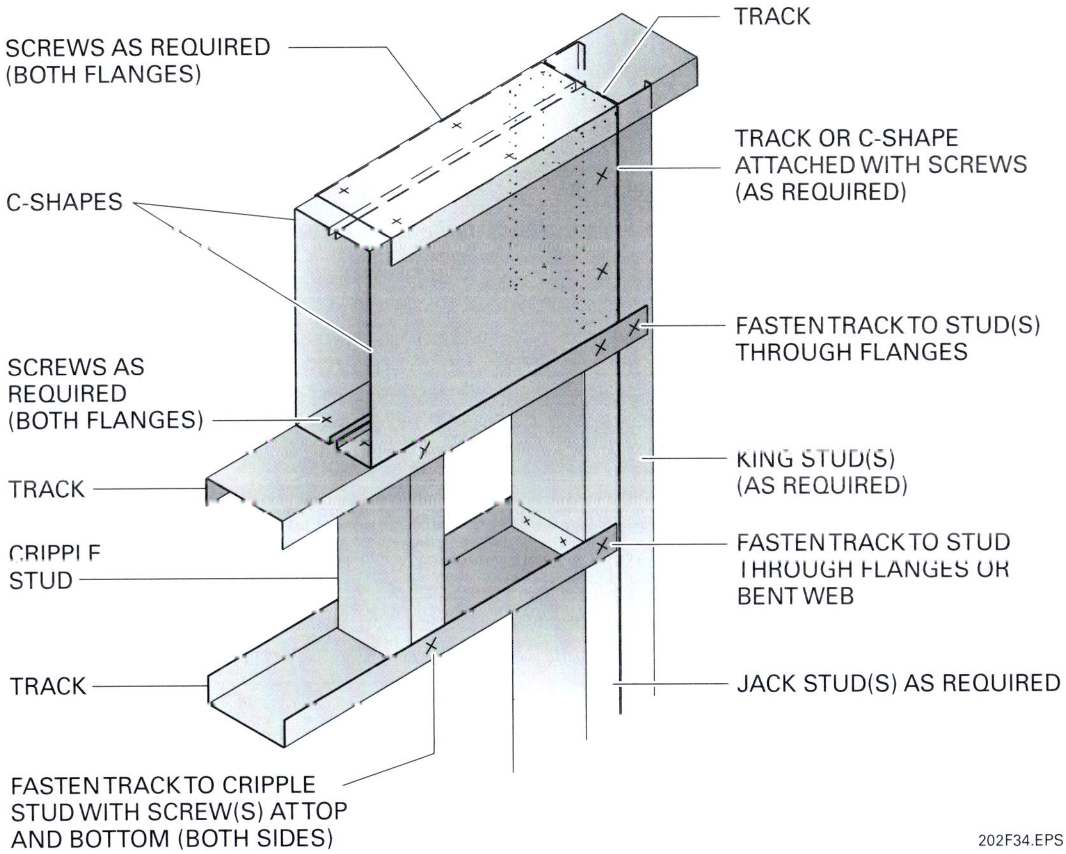

202F34.EPS

Figure 34 ♦ Loadbearing header.

be ordered from manufacturers. Selection of the header type depends on loads and applications. L-headers are being used more frequently in low-rise and multi-family construction because they use fewer fasteners and less material.

To install the box header, loosen one of the king studs at the top of the rough opening, and install the header with the open end up, by inserting it into the top track. This is usually a tight fit. Clamp the header at one end with a bar clamp. Apply pressure, working from one side of the header to the other, using the clamp to tightly push the header into the top track. Make sure that the header is fitted tightly into the track before screwing it in place. Reposition the king studs that were loosened, and screw them back into place.

Back-to-back headers are formed by placing two C-shapes with the webs of the members touching each other. They are positioned in the top track of the wall and finished just like a box header.

The L-header (*Figure 35*) consists of one or two angle pieces that fit over the top track. This design saves labor because no special fabricating is needed, and the number of screws is reduced. The L-shape itself spans the opening for the header.

Once the header has been installed in the top track, the remainder of the rough opening may be framed (*Figure 36*).

NOTE

After header assembly, the window opening on the drawing should be rechecked to verify the position and height of the window in the wall.

Mark the jack (trimmer) studs for the top of the window opening. Another mark should be made on the trimmer studs at the bottom of the window. Cut two track pieces to fit between the jack studs to make up the head and sill pieces. These pieces should be cut 2 inches longer than the distance between the jack studs so the flanges can be clipped 1 inch on each side.

Clip the flanges and bend the web down toward the flanges. Set the head and sill pieces in the opening, keeping the hard side of the track towards the opening. Screw the tab at the flange of each piece into the jack studs with one No. 8 screw at each tab. The tabs on the webs of the header track should also be screwed into the trimmer studs with two No. 8 screws.

202F35.EPS

Figure 35 ◆ L-header.

202F36.EPS

Figure 36 ◆ Completed rough opening.

The cripple studs should be cut to fit between the header track and head pieces, as well as between the sill piece and bottom track. The cripples should maintain the spacing layout (16 or 24 inches on center) for ease in installing gypsum board and sheathing. Screw the cripple studs into place with a No. 8 screw at each track flange on both sides.

Door openings do not require a bottom sill. However, the bottom track should run continuously at the bottom of the door to hold the wall together temporarily. The track can be cut out after the wall is plumb, level, and permanently braced.

5.1.6 Jambs

Loadbearing jambs require a minimum of two studs on each side of the framed opening (one trimmer plus one king). More studs will be needed if they are required by the building code or by engineering analysis. The studs must be fastened together to act as one member, either by capping them with track and screw fastenings or by stitch welding (*Figure 37*).

5.1.7 Sills

Sill members are usually single-track sections that are the same width as the wall stud used to frame an opening. They are clipped and screwed to the jamb studs. In cases where the allowable lateral load is exceeded, sills must be constructed with multiple track sections.

5.2.0 Steel Floor Assemblies

Cold-formed steel floor assemblies typically use standard C-shape floor joists (*Figure 38*), proprietary floor joists, pre-engineered steel floor trusses, **rim track**, web stiffeners, **clip angles**, hold-down anchors, and fasteners. They can be installed on crawl spaces and stem walls, as well as directly to interior structural walls. Steel floor assemblies are similar to conventional framing and use single- or multiple-span installation techniques.

5.3.0 Steel Roof Assemblies

Recent years have seen a dramatic rise in the use of cold-formed steel framing members in roof assemblies. It allows for easy and standardized assembly, and they are durable and non-combustible. In addition to the standard C-shaped member, scores of proprietary shapes, fabrication methods and installation requirements are available from truss manufacturers nationwide. *Figure 39* shows an example of custom-designed steel trusses.

202F37.EPS

Figure 37 ◆ Stitch-welded jamb.

202F38.EPS

Figure 38 ◆ Standard floor joists.

202F39.EPS

Figure 39 ◆ Example of complex framing using steel trusses.

Trusses are made from structural members that have been connected into triangular sections. Multiple trusses are used to make up the structural framework of a roof. Since all trusses must be constructed to meet strict standards, they are often premanufactured rather than constructed in the field. When trusses must be built in the field, always measure and cut members accurately and connect them according to the job specifications.

6.0.0 ♦ BRACING STEEL WALLS

Bracing prevents a cold-formed steel framing member from twisting and buckling. In wall construction, there are several common methods for bracing, and all have different applications and purposes. *Figures 40* through *42* are examples of typical types of bracing.

Intermediate stud bracing – This bracing is used when gypsum board or structural sheathing is not applied to both sides of loadbearing walls, such as garage walls (*Figure 43*).

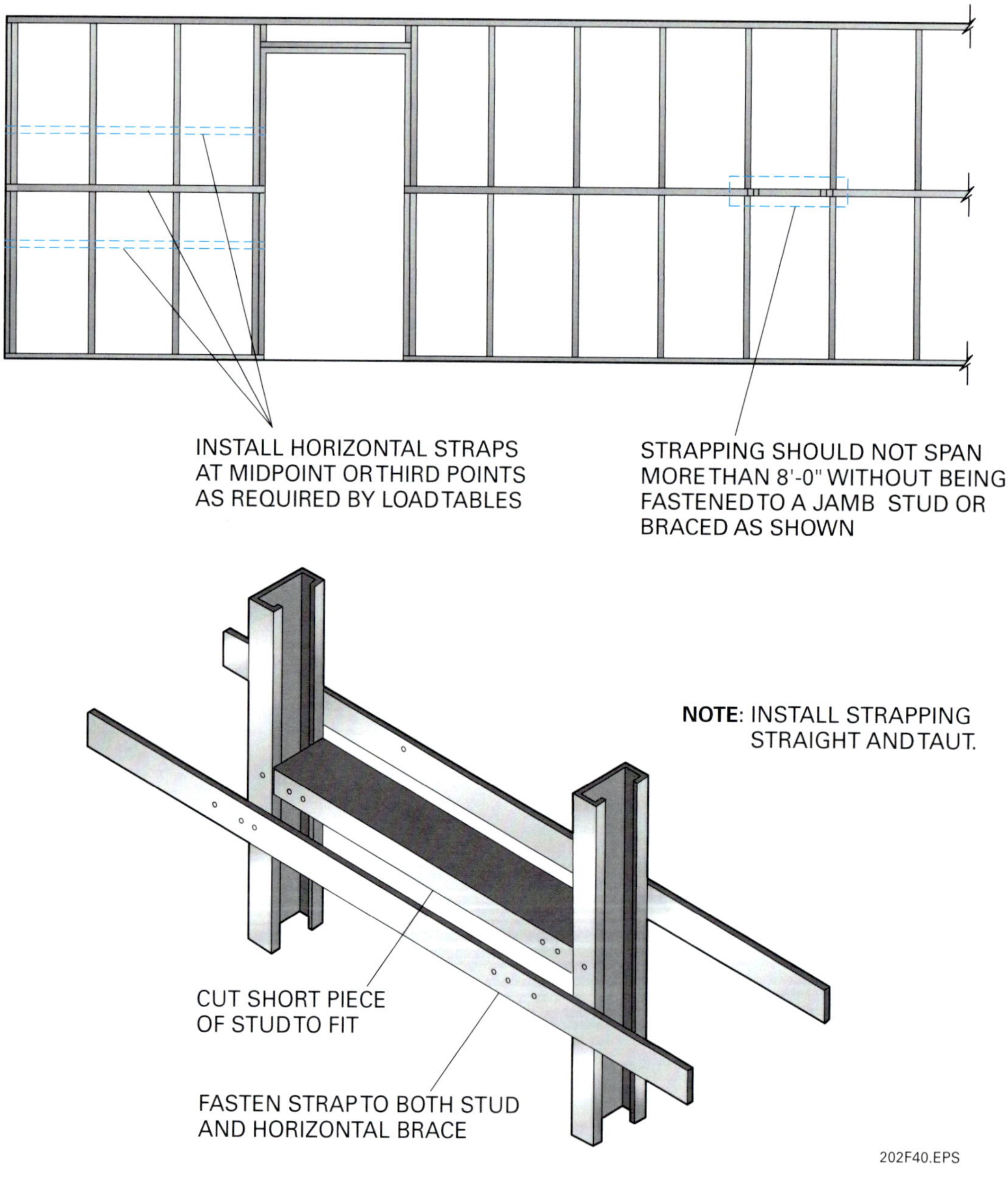

Figure 40 ♦ Lateral strapping for stud walls.

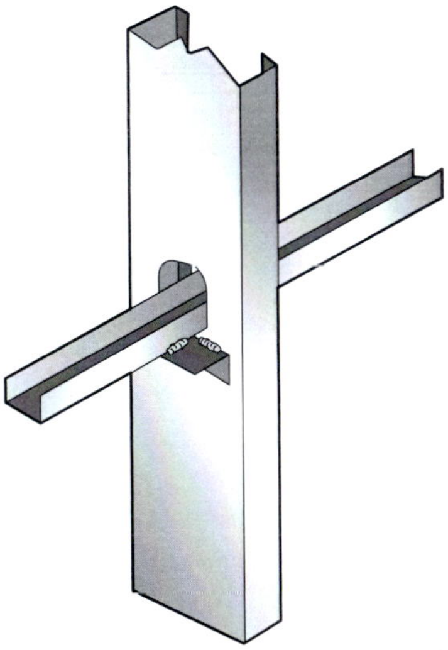

Figure 41 ◆ Bridging with cold-rolled channel (CRC).

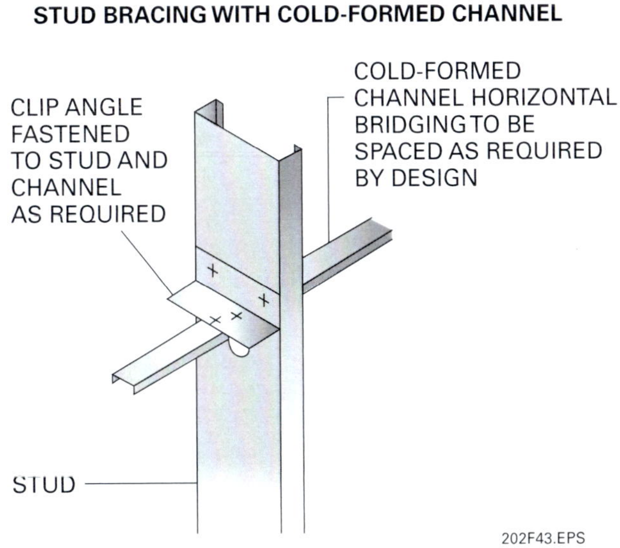

Figure 43 ◆ Example of stud bracing.

NOTES:

- Install strapping as close to 45° as possible.
- Place straps on both sides of stud wall in order to prevent eccentric loading.
- Check for increased axial load applied to wall studs due to tension in straps. Double studs will typically be required at strap ends.
- Install wall straps straight and taut.

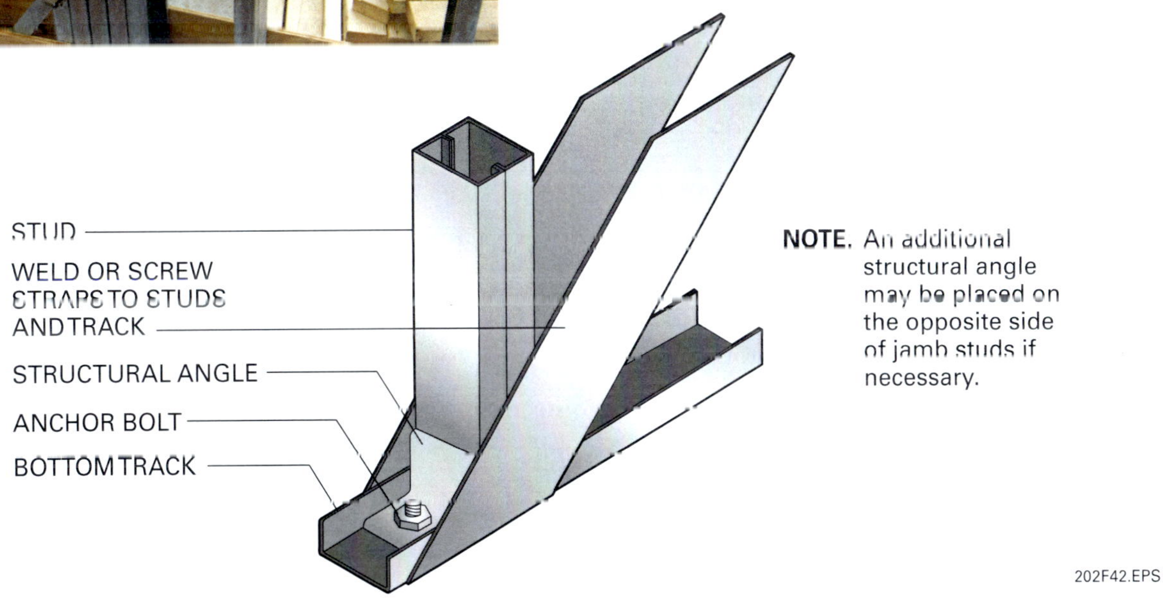

Figure 42 ◆ Diagonal strapping for shear walls.

Shear wall bracing – There are two ways of applying shear wall bracing: structural sheathing (*Figure 44*) and X-bracing (*Figure 45*). Structural sheathing Type II plywood or OSB may be adequate to keep the wall from **racking** depending on the design, as long as there are not excessive openings in the wall. In order for structural sheathing to be effective, it should be installed with the long dimension parallel to the stud framing (vertical orientation). The plywood may be secured to the wall while panelizing, or after the wall is plumb and level. X-bracing is another way to obtain shear strength when structural sheathing is not used. X-braces are diagonal steel straps attached to the walls with screws or welded connections. This bracing must be designed by an engineer, and the straps must be inspected for the correct number of fasteners. Do not tighten the straps until the walls are plumbed and aligned.

Temporary bracing – There are two types of basic temporary bracing: one for panelized walls and one for installed walls. After a straight wall is constructed on a panel table, installing plywood or temporary bracing prevents racking when the wall is removed from the table. Before the wall is taken off the table, check for squareness by diagonally measuring the panel. When both diagonals are equal, the wall is square. Adjust if necessary. Lay extra studs or truss material across the wall diagonally, and screw the bracing to the wall studs, especially at door openings where the bottom track is weak. Leave the bracing on the wall until the wall is installed and permanently braced. These precautions will help provide straight walls that are ready for final installation.

Installation of steel-framed walls requires adequate temporary bracing to resist loads during construction until permanent bracing can be installed (*Figure 46*).

202F44.EPS

Figure 44 ◆ Wood structural sheathing.

202F45.EPS

Figure 45 ◆ Gusset plate X-bracing.

7.0.0 ◆ NONSTRUCTURAL (NONBEARING) WALL FRAMING

Framing members used in interior systems may be nonbearing or may be designed as part of the structural system. This section primarily covers nonbearing walls that function as space partitions within the exterior walls of a building.

202F46.EPS

Figure 46 ◆ Temporary bracing.

Similar to structural wall systems, nonstructural walls are comprised of studs, track, and accessories. The primary differences are the characteristics of the materials and the application of connectors and accessories.

Nonstructural framing members typically have a base steel thickness of 18, 27, or 30 mils, compared with a minimum thickness of 33 mils for structural studs. In addition, the minimum stud flange dimension for nonstructural framing members is 1.25 inches, and the minimum return (stiffening) lip dimension is ⅛ inch, compared with 1⅝ inch flange and ½ inch lip for structural studs. Also, nonstructural members typically have a G40 galvanized coating weight, compared with G60 or higher for structural studs. The rules for nonstructural framing are different from those for structural walls. These rules may be found in the gypsum specifications, rather than in the building codes.

Since nonstructural studs are not intended to carry any loads, accessories used in this application are used for architectural features within the building, including curves and arches. No connection is required between the stud and track for nonbearing walls, except for studs that are adjacent to window and door openings.

7.1.0 Steel Curtain Walls

Since their introduction in the early 1900s, metal and glass curtain-wall systems have become very popular in the architectural design of contemporary structures. Unlike interior partitions, wind-bearing curtain walls resist loads from exterior wind pressures that, in some cases, exceed 60 pounds per square foot. Cold-formed steel curtain walls are made up of various components, which may include the following:

- *Angles* – Angles are used to connect framing members within the curtain wall system.
- *Clip angle* – A steel angle, generally 3 to 12 inches long, that makes the transition between a framing member and the component supporting it. These angles are used to connect two framing members.
- *Continuous angle* – A steel angle that makes the transition between a stud curtain wall and the primary frame. The angle is typically of hot-rolled thickness (³⁄₁₆ to ⅜ inch); however, thinner-gauged materials can be used if the span and load requirements are relatively small.
- *Diagonal brace (or kicker)* – A sloping brace used to provide lateral support to a curtain wall assembly. When installed horizontally, a diagonal brace is referred to as a strut.
- *Embed* – A hot rolled steel plate or angle, reinforced with shear studs or steel rebar, that is cast into a concrete floor or beam. Embeds allow for the welded attachment of steel supports.
- *Girts* – Horizontal structural members that support wall panels and are primarily subject to bending under horizontal loads, such as wind load.
- *Slide clip* – A connection device that permits the primary frame to which a stud attaches to move axially, while it braces the stud against lateral forces.
- *Slip track* – A track section used in for-fill curtain wall applications (*Figure 47*). Slip tracks accommodate vertical movements of a primary frame (normally ¼" to ¾"), while bracing the wall against lateral forces. Slip tracks may also be specified at the top of interior drywall partitions.

The term *curtain wall* is used to distinguish this system from loadbearing framing. As described earlier in this section, loadbearing framing requires that the wall members carry the weight of the structure that is above them. With curtain-wall framing, the structure is usually already in place, and the wall framing is filled-in between the floor slabs. The stud-to-track gap distance for curtain wall is no more than ¼ inch, unless it is otherwise specified in an approved design. This is different from loadbearing construction, which permits only a ⅛-inch gap. The only gravity loads

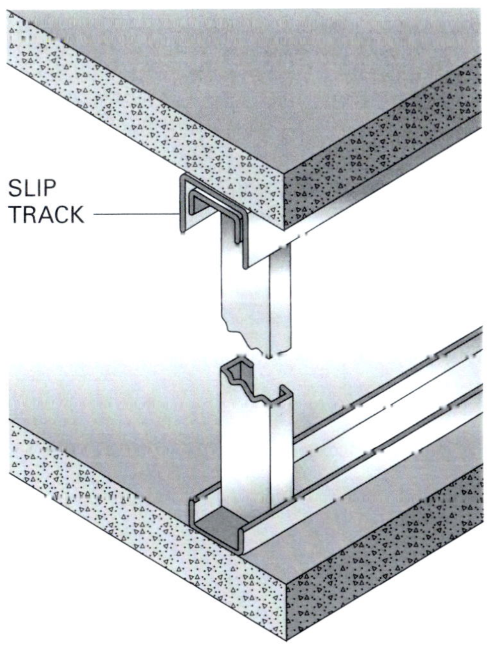

Figure 47 ♦ Slip track.

that curtain wall framing typically carry are the weight of any cladding or finish materials attached to it.

CAUTION

For curtain wall installation, the use of components formed from steel measuring less than 33 mils should be avoided.

7.1.1 Construction Methods for Steel-Framed Curtain Walls

The following are four main methods for building assemblies in curtain wall construction:

- *Infill* – This term describes applications in which studs are only one story tall, spanning from floor to floor of a structure. Infill framing requires less stud material. Connections are often easier to make, since low-cost powder-actuated pins can be used in many applications. In addition, the spans are often shorter than bypass conditions, so thinner steel or wider spacing can be used. The disadvantage of infill framing is that more track material is needed, and wall sections can be difficult to panelize.
- *Panelization* – This method can be used if field measurements are made after the floor systems are in place, or if a slip connector or telescoping stud system is used. When structural steel is used for main framing, the **spandrel** beam can get in the way of framing. In that case, framing must be attached to the bottom of the beam. If framing takes place outside the spandrel beam, it can cause insulation support difficulties.
- *Bypass framing (balloon framing)* – This technique allows a single stud to be used for framing two or more floors, and it requires less track material than other types of framing. The multiple spans can reduce moment stresses in members, allowing for more widely spaced or thinner framing members. In addition, this method makes it easier to pre-panelize large sections, including multi-story panels. On the other hand, some connections can be more difficult to make, such as bracing and support connections behind columns and spandrel beams. Depending on the conditions, double connections may need to be made. This requires that clips or slip connectors are attached first to the structure and then again to the stud. This last issue can be lessened somewhat by the use of connectors that friction-fit inside the stud, thus reducing connection time. Finally, if slab or other structural elements extend into the stud cavity, they may need to be chipped away, or stud framing may need to be altered to correctly install the bypass framing.
- *Stacked wall framing* – This method permits bypass framing while isolating slip connections to one- or two-story segments. As a result, multi-story panels and pre-finished panels can be fabricated and installed, including insulation. However, to prevent water infiltration, this approach requires special detailing at slip connections and exterior finishes. Because the entire panel weight goes to fixed connections, usually at every other floor, these fixed connections need to be strengthened to carry the added dead load of the taller panels. Also, some connections are more difficult to reach, such as bracing and support connections behind columns and spandrel beams.

7.1.2 Bracing for Curtain Walls

The most common cold-formed steel curtain wall shape is the C-stud. Due to its shape and geometric properties, a C-stud tends to rotate under lateral load. Nonbraced studs may also move out of plane. This is known as torsional-flexural buckling. Mechanical bridging and/or sheathing materials can restrain the flanges and prevent this condition. When using discreet bracing rather than sheathing, decreasing the spacing of the bridging typically increases the member capacities. Increasing the spacing decreases capacities.

In a typical stud-framed exterior curtain wall, member deflections are the primary serviceability issue. Deflection limits are most often determined by the architectural finishes and are set forth in the project specifications.

7.1.3 Finishing for Curtain Walls

A wide range of finish systems may be applied to steel stud curtain-wall frames, including the following:

- Brick veneer
- Split-faced block veneer
- Tile or thin-cast brick
- Exterior insulation finish systems (EIFS)
- Glass fiber reinforced concrete (GFRC)
- Metal panel
- Modified portland cement (stucco)
- Fiber-cement board or siding
- Dimensional stone, such as granite or limestone
- Wood siding and many other finish systems

7.2.0 Radius (Curved) Walls

Some partitions that might be difficult to construct with wood framing members can easily be built with steel framing members. A radius wall is an example. Note that the wall forms a quarter circle. Constructing a wall of this radius with wood framing members would take many hours. Using steel framing members, the wall could be built in less than an hour. A plywood template may be required for more complex radius walls.

The bend can be achieved using several methods. Curved walls can be framed out of steel using curved track for partitions or exterior loadbearing walls. The track can be bent at the job site by slitting the flanges. Ordering curved track from specialty companies also speeds up the construction of a radius wall.

Track may also be ordered curved to a specified radius. Some specialty companies use machinery to bend the track without slitting the flanges; others produce a flexible track that can be ordered and formed on the job site based on the desired effect. This provides a clean, neatly bent track to an exact radius. Wall track is bent around the flanges (*Figure 48*).

7.3.0 Other Nonstructural Wall Assemblies

Some wall assemblies are nonstructural. This means that they are not loadbearing and are not needed to maintain the structure. Examples would be a partial partition or a chase wall designed to conceal wiring or plumbing.

202F48.EPS

Figure 48 ◆ Example of a radius track.

- *Chase walls* – It is common in commercial and residential construction to build chase walls to conceal plumbing and other utilities. For oversized utilities or acoustical requirements, two separate walls with a void in between may be framed in advance. Cross-bracing between the two walls may be required.
- *Fire-rated assemblies* – Building codes require that certain partitions provide fire-rated separation of interior spaces, including a specific rating that states how long the partition or assembly would last under the design load before being penetrated by a fire. Approved assemblies must be constructed precisely as described in the directories published by the rating agencies.
- *Area separation walls (zone walls)* – This term refers to a specific group of rated walls, designated by code requirements, which are designed to permit structural failure on one side of the wall while still providing fire protection.

INSIDE TRACK Curves

One of the most beneficial features of steel framing is that it can be easily bent into curves. This allows for curved walls and dome-shaped roofs. Steel members may be purchased already bent into standard shapes, or they may be bent manually. Manual bending, however, is a tedious, time-consuming job that requires lots of measuring. Another option is to use a track bender tool. This device can be used on site to quickly and accurately bend tracks to the desired shape without measuring by simply setting a dial.

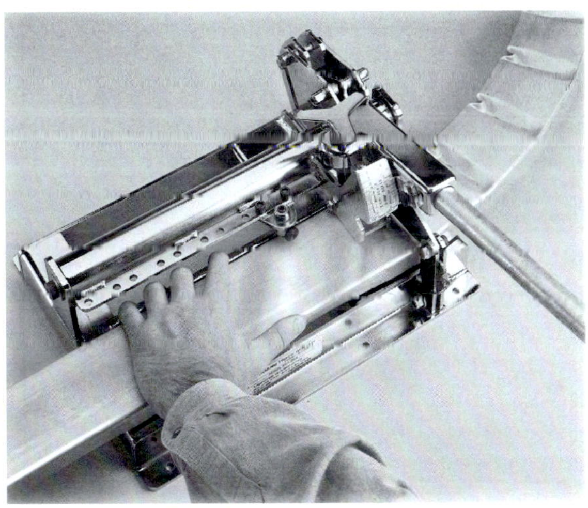

202SA02.EPS

- *Head-of-wall conditions* – Some building codes require that wall assemblies accommodate horizontal and vertical movement while still providing a fire-resistive barrier. Proprietary and nonproprietary devices and assemblies have been designed and tested to meet code requirements.
- *Shaftwall systems* – These structures are special types of rated wall systems that may be constructed from one side only. Higher sound and fire ratings may be achieved by attaching additional layers of gypsum board to the outer or inner face. A 1-inch core board is specifically made for this purpose. Horizontal shaftwall may also be used for soffits or **plenums** where ratings are required.

7.4.0 Furring

Design requirements may specify the standoff of gypsum board materials from the stud using hat-shaped steel furring members. This provides sound isolation. The gypsum board is usually furred out using resilient channel (*Figure 49*). Resilient furring channels are used over both metal and wood framing to provide a sound-absorbent spring mounting for gypsum board. The channels should be attached as specified by the manufacturer. They not only improve sound insulation, but also help isolate the gypsum board from structural movement, minimizing the possibility of cracking. Resilient furring channels may also be used for application of gypsum boards over masonry and concrete walls. In wood frame construction, gypsum board can be screwed to resilient metal furring channels to provide a higher degree of sound control. Furring has additional applications as follows:

- Provides additional space for insulation
- Allows out-of-plane walls or walls of different thickness to match and have a smooth surface
- Provides additional space to conceal fixtures or structural elements within a wall

7.4.1 Ceiling Systems

Ceiling systems usually consist of a suspended drywall grid with gypsum board. An alternative method is to use furring and cold-rolled channel to provide a rigid framework for suspended gypsum board and other ceiling assemblies, with the steel members suspended (*Figure 50*). The furring channel is perpendicular to the underside of the U-channel and is commonly clipped or wire-tied at appropriate intervals for attaching gypsum board with screws.

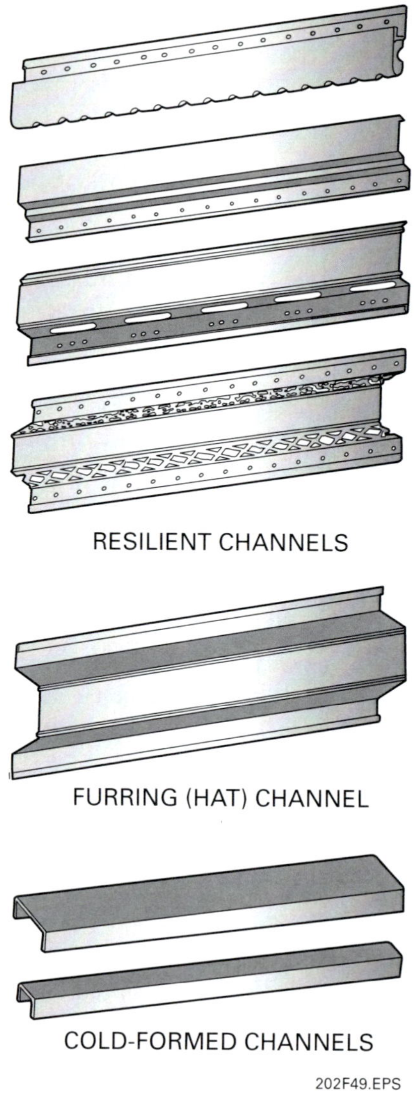

Figure 49 ◆ Types of furring channels.

Figure 50 ◆ Ceiling framework.

The furring channel can also be screwed to other structural steel members, and can be attached directly to the bottoms of bar joists. In the latter case, the furring is installed perpendicular to the joists and wire-tied at appropriate intervals. Note that wire ties are required for fire-rated and multi-layer assemblies.

8.0.0 ◆ SLIP CONNECTIONS

Slip connectors (*Figure 51*) are devices that allow for the vertical movement of a structure without imposing additional loads on cold-formed steel framing or other wall components. Slip connectors are used when a structural system other than steel framing is needed to carry the loads of upper floors and the roof down to the foundation. Loads on the upper parts of a structure cause it to deflect, which in turn may induce loading in walls and wall components that are not designed to carry these loads. Slip connectors are designed to allow for this movement without creating unmanageable stress on components.

Slip connectors are generally located at the top of a wall panel, where it meets the underside of a structural element, such as a floor slab or beam. Under gravity, seismic, or wind loads, this upper portion of the structure may deflect up or down. The connector is designed to allow this movement, restrain the wall system from out-of-plane movement, and prevent any additional axial loading on the stud.

Slip connections are also useful at locations where a wall system is continuous, bypassing intermediate floors. When this occurs, a slip connection extends from the side of the structure and supports the wall components laterally. Connections are also installed at roof bypasses where the wall system extends past a roof structure to form a parapet or high wall. At such locations, the slip connection must permit movement of the structure either up, due to wind uplift, or down, due to gravity loads.

In multi-story construction, movement may occur at the floor below a curtain wall system, causing the entire wall to move down. When this happens, the connector at the top of the wall (either a bypass or below structure) must have sufficient capacity to allow the wall system to move down without creating tension on wall components.

202F51.EPS

Figure 51 ◆ Slip connectors.

8.1.0 Other Typical Connections

Other than slip connections, the most typical connections used in stud curtain-wall design include the following:

- *Base connections* – These connections may be stud to track, track to concrete, or steel deck.
- *Head or sill-to-jamb connections* – Points where the head or sill is directly connected to the jamb.
- *Continuous angle connections* – A field-welded connection usually found at spandrel framing.
- *Clip angle connections* – Typically found at non-slipped bypass or spandrel framing connections; designed to carry both lateral and self weight forces.
- *Outrigger clips* – Short lengths of angles designed as an axially-loaded strut to carry stud lateral reactions back to the structure.
- *Wind girt connections* – Usually located at taller spandrel framing cases when the use of diagonal braces is impractical. These connections are not usually required to carry any gravity loads.
- *Diagonal braces or kickers (as opposed to X-bracing for shear walls)* – These stud diagonals provide for a bottom spandrel stud reaction back to the structure.
- *Stud-to-stud connections* – May be either lapped or track-to-track. These connections occasionally require movement allowance, which can be accommodated by a slip-track or slip-pin type detail.
- *Knee wall base* – This is a moment connection at the bottom of a knee or stub-wall; is usually either a freestanding parapet or a long segment of wall under a continuous or ribbon window condition.

- *Deep leg track* – Also known as a deflection track. The deep leg track allows for some degree of vertical deflection of the wall, without the wall sustaining damage due to lateral movement. In this type of connection, the tops of the studs are not secured to the upper track. When the upper part of the structure is under load, the track moves down until it rests on the tops of the studs. When this method is used, the studs must be properly stabilized to prevent twisting and bending.

INSIDE TRACK

Design Software

Designing a steel frame structure requires a great deal of knowledge about structural design and math. Until recently, designers manually computed dimensions, weights, and live loads. Today, computers help to make designing steel-framed structures easier and faster. Software is available that assists designers in planning a steel-framed structure. One such application is called The Steel Smart System®. It is specifically designed to be used with components designed by The Steel Network. The sample screen shown allows the operator to design a wall with a window opening. The operator fills in information, such as the size of the steel members, loads applied to the members, anticipated horizontal and vertical deflections, stud spacing, and bracing. When all data has been supplied, it is processed to determine if the design is safe. While this type of software does not take the place of an engineer, it does help engineers to do their jobs.

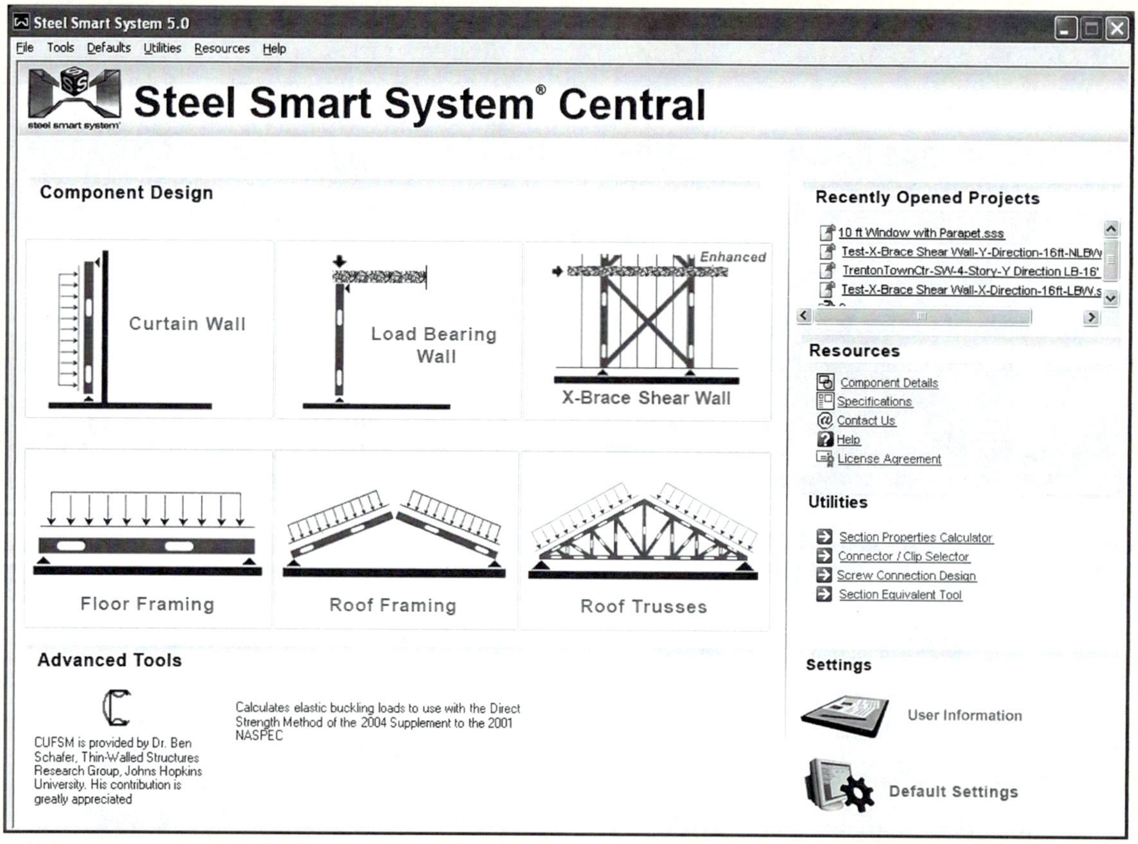

202SA03.EPS

Review Questions

1. The maximum speed of screwguns for steel-to-steel screw connections is _____ rpm.
   a. 1,500
   b. 2,000
   c. 2,500
   d. 3,000

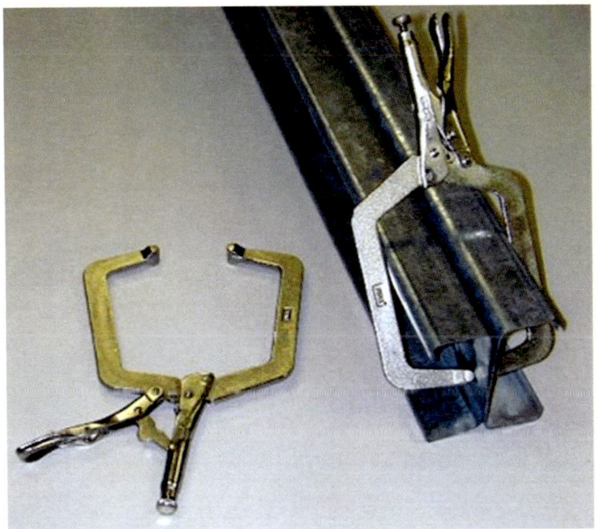

202RQ01.EPS

Figure 1

2. The tool shown in *Figure 1* is a(n) _____.
   a. channel stud shear
   b. locking C-clamp
   c. aviation snip
   d. stud crimper

3. A steel framing component that has a web, two flanges, and two lips is called a _____.
   a. cripple stud
   b. C-shape
   c. web stiffener
   d. header

4. A cutting tool that can cut steel components up to 33 mils thick and leave no abrasive edges is a _____.
   a. chop saw
   b. snip
   c. C-clamp
   d. swivel-head shear

5. One mil is equivalent to _____ inch.
   a. 1/10
   b. 1/100
   c. 1/1,000
   d. 1/10,000

6. When selecting a screw to attach steel framing parts, the screw length should be the thickness of all the material plus _____ inch.
   a. 1/8
   b. 1/4
   c. 3/8
   d. 1/2

7. When using bolts to fasten steel framing to concrete, you must install _____.
   a. a washer
   b. a hex washer head screw
   c. caulking
   d. pneumatic pins

8. The organization that created the standard designation system for steel components is the _____.
   a. American Iron and Steel Institute
   b. OSHA
   c. United Steel Alliance
   d. Steel Stud Manufacturers Association

9. In the STUFL system, the letter L represents a(n) _____.
   a. furring channel
   b. track
   c. stud with lips
   d. angle

10. The part of a C-shape framing member that extends from the edge of a flange as a stiffening element is called a _____.
    a. lateral
    b. lip
    c. web
    d. knurl

11. The maximum center line to center line tolerance for in-line framing of cold-formed steel is _____ inch.
    a. 1/8
    b. 1/4
    c. 1/2
    d. 3/4

Review Questions

12 Structural stud sizes are typically _____ inches in width.

a. 1 to 6
b. 1⅝ to 6
c. 2 to 5
d. 2½ to 8

13. A wall that is designed to resist lateral forces such as those caused by earthquakes or wind is called a _____ wall.

a. curtain
b. panel
c. shear
d. radius

14. Diagonal steel straps attached to the walls are used to _____.

a. brace shear walls
b. temporarily brace panelized walls
c. provide structural sheathing
d. support headers

15. The normal thickness of nonstructural framing members is from _____ mils.

a. 12 to 18
b. 18 to 30
c. 30 to 33
d. 33 to 47

Summary

Cold-formed steel framing is much more widespread in the construction industry than it was a few years ago. Steel will not warp, swell, or split. It is immune to termites and will not contribute to a fire. It is a perfect fit for noncombustible construction and environmentally conscious building, and steel is more cost-effective than lumber. In a wide variety of residential and commercial buildings, cold-formed steel is being used not just for partition walls but as the main structural component. While the layout of cold-formed steel framing members may be very similar to conventional wood framing, there are important differences to remember in components and techniques. The studs, fasteners, and various accessories are designed to take full advantage of the light weight and design flexibility of steel.

Notes

Trade Terms Introduced in This Module

Blocking: C-shaped track, break shape, or flat strap material attached to structural members, flat strap or sheathing panels to transfer shear forces.

C-shape: A cold-formed steel shape used for structural and nonstructural framing members consisting of a web, two flanges, and two lips (edge stiffeners).

Clip angle: An L-shaped piece of steel (normally having a 90-degree bend), typically used for connections.

Cold-formed steel: Sheet steel or strip steel that is manufactured by press braking blanks that have been sheared from sheets or cut length of coils or plates, or by continuous roll forming of cold- or hot-rolled coils of sheet steel.

Curtain wall: A light, nonbearing exterior wall attached to the concrete or steel structure of a building.

Diaphragm: A floor, ceiling, or roof assembly designed to resist in-plane forces such as wind or seismic loads.

For-fill: A wall to be filled with some type of material, such as concrete or insulation.

Header: A horizontal structural framing member used over floor, roof, or wall openings to transfer loads around the opening to supporting structural framing members.

Knurled: A series of small ridges on pins used to provide a better gripping surface on metal and plastic.

Lateral: Running side to side; horizontal.

Lips: That part of a C-shape framing member that extends from the flange as a stiffening element that extends perpendicular to the flange; also called edge stiffener.

Mil: A unit of measurement equal to $\frac{1}{1,000}$ of an inch.

Panelization: The process of assembling steel-framed walls, joists, or trusses before they are installed in a structure. Roll-formers normally cut studs to within $\frac{1}{8}$-inch tolerance. This helps the framer to consistently create straight walls on the panel table that are easy to install in the field.

Plenum: A confined space such as the area between a suspended ceiling and an overhead deck that is used as a return for a heating, cooling, and ventilation systems.

Racking: Being forced out of plumb by wind or seismic forces.

Rim track: A horizontal structural member that is connected to the end of a floor joist.

Shear wall: A wall designed to resist lateral forces such as those caused by earthquakes or wind.

Spandrel: A beam that is an angled support for framing.

Track: A framing member consisting of only a web and two flanges. Track web depth measurements are taken to the inside of the flanges.

Appendix A

Common Terms Used in Cold-Rolled Steel Framing Work

AISC: American Institute of Steel Construction.

AISI: American Iron and Steel Institute.

Applicable building code: Building code under which the building is designed.

Approved: Approved by a building official or design professional.

Base steel thickness: The thickness of bare steel exclusive of all coatings.

Bracing: Structural elements that are installed to provide restraint or support (or both) to other framing members so that the complete assembly forms a stable structure.

Ceiling joist: A horizontal structural framing member that supports ceiling components and which may be subject to attic loads.

Cold-formed sheet steel: Sheet steel or strip steel that is manufactured by press braking blanks sheared from sheets or cut length of coils or plates, or by continuous roll forming of cold- or hot-rolled coils of sheet steel. Both forming operations are performed at ambient room temperature; that is, without any addition of heat such as would be required for hot forming.

Cold-formed steel: See *cold-formed sheet steel.*

Cold-formed steel structural framing: The elements of the structural frame, as given in the *Code of Standard Practice, Section B1.*

Component assembly: A fabricated assembly of cold-formed steel structural members that is made by the component manufacturer; may also include structural steel framing, sheathing, insulation, or other products.

Component design drawing: The written, graphic and pictorial definition of an individual component assembly, which includes engineering design data.

Component designer: The individual or organization responsible for the engineering design of component assemblies.

Component manufacturer: The individual or organization in charge of the manufacturing of component assemblies for the project.

Component placement diagram: The illustration, supplied by the component manufacturer, identifying the assumed location for each of the component assemblies and referencing each individually designated component design drawing.

Construction manager: The individual or organization selected by the owner to issue contracts for the construction of the project and to purchase products.

Contract documents: The documents including, but not limited to, plans and specifications that define the responsibilities of the parties involved in bidding, purchasing, designing, supplying, and installing cold-formed steel framing.

Contractor: The individual or organization that is contracted with to assume full responsibility for the construction of the structure.

Cripple stud: A stud that is placed between a header and a window or door head track, a header and wall top track, or a window sill and a bottom track to provide a backing to attach finishing and sheathing material.

Design professional: A person who is registered or licensed to practice a design profession as defined by the statutory requirements of the state in which the project is to be constructed.

Design thickness: The steel thickness used in design which is equal to the minimum base steel thickness divided by 0.95.

Drawings: See *plans* and *installation drawings.*

Edge stiffener: That part of a C-shape framing member that extends from the flange as a stiffening element that extends perpendicular to the flange.

Erection drawings: See *installation drawings.*

Erector: See *installer.*

Flange: That portion of the C-shape framing member or track that is perpendicular to the web.

Floor joist: A horizontal structural framing member that supports floor loads and superimposed vertical loads.

Framing contractor: See *installer.*

Framing material: Steel products, including but not limited to, structural members and prefabricated structural assemblies, ordered expressly for the requirements of the project.

General contractor: See *installer.*

Harsh environments: Coastal areas where additional corrosion protection may be necessary.

In-line framing: Framing method in which all vertical and horizontal load-carrying members are aligned.

Installation drawings: Field illustrations that show the location and placement of the cold-formed steel structural framing.

Installer: Party responsible for the installation of cold-formed steel products.

Jack stud: A stud that does not span the full height of the wall and provides bearing for headers. Also called a *trimmer stud.*

King stud: A stud, adjacent to a jack stud, that spans the full height of the wall and supports vertical and lateral loads.

Lip: See *edge stiffener.*

Material supplier: An individual or entity responsible for furnishing framing materials for the project.

Nonstructural member: A member in a steel-framed assembly which is limited to a transverse load of not more than 10 lb/ft^2 (480 Pa); a superimposed axial load, exclusive of sheathing materials, of not more than 100 lb/ft (1,460 N/m); or a superimposed axial load of not more than 200 lbs (890 N).

Owner: The individual or entity organizing and financing the design and construction of the project.

Plans: Drawings prepared by the design professional for the owner of the project. The drawings include, but are not limited to, floor plans, framing plans, elevations, sections, details, and schedules as necessary to define the desired construction.

Punchout: A hole made in the web of a steel framing member during the manufacturing process.

Shop drawings: Illustrations for the production of individual component assemblies for the project.

Span: The clear, horizontal distance between bearing supports.

Specifications: Written instructions that, with the plans, define the materials, standards, product designs, and workmanship expected on a construction project.

Standard cold-formed steel structural shapes: Cold-formed steel structural members that meet the requirements of the SSMA Product Technical Information.

Strap: Flat or coil sheet steel material, typically used for bracing and blocking, that transfers loads by tension and/or shear.

Structural engineer-of-record: The design professional responsible for sealing the contract documents indicating that he or she has performed or supervised the analysis, design, and document preparation for the structure and has knowledge of the requirements for the load-carrying structural system.

Structural member: A member that is designed or intended to carry loads, such as a floor joist, rim track, structural stud, wall track in a structural wall, ceiling joist, roof rafter, or header.

Structural stud: A stud in an exterior wall, or an interior stud which supports superimposed vertical loads and which may transfer lateral loads, including full-height wall studs, king studs, jack studs, and cripple studs.

Stud: A vertical framing member in a wall system or assembly.

Trimmer: See *jack stud.*

Truss: A coplanar system of structural members joined together at their ends, usually to construct a series of triangles that form a stable, beam-like framework.

Yield strength: A characteristic of the basic strength of the steel material defined as the highest unit stress that the material can endure before permanent deformation occurs as measured by a tensile test in accordance with *ASTM A 370.*

Appendix B

Interpreting STUFL

Need the STUFL? Get it here…

The Right STUFL:

Universal Designator System for Cold-Formed Steel Framing Members

The Right STUFL will identify any common cold-formed steel framing members using:

Web Depth (D), expressed in 1/100th inches,
Flange Width (B), expressed in 1/100th inches,
Minimum Base Steel Thickness, expressed in mils (1/1000th inches), and the following designators,

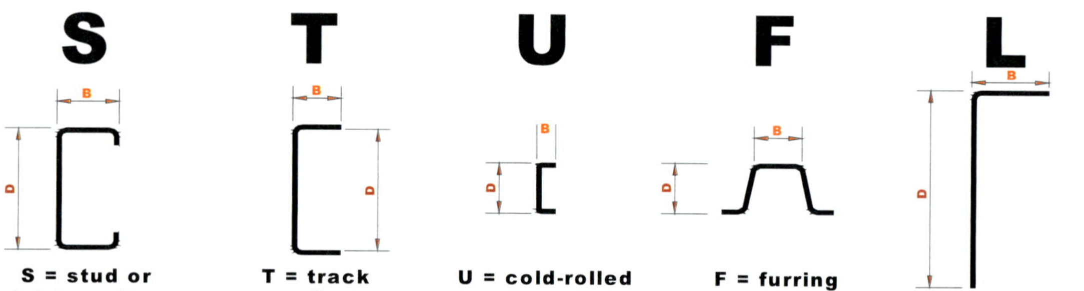

S = stud or joist section with flange stiffeners

T = track sections

U = cold-rolled channel

F = furring channel

L = L-header or angle

Examples

Designation for a 5-1/2" – 16 gauge C-shape with 1-5/8" flanges: 550S162-54

550 S 162 —54

Minimum base steel thickness in mils (.054 in = 54 mils)
1-5/8" flange in 1/100th inches
Stud or joist with flange stiffeners
5-1/2" member depth in 1/100th inches (outside to outside dimension)

Designation for a 3-1/2" – 20 gauge Track with 1-1/4" flanges: 350T125-33

350 T 125 —33

Minimum base steel thickness in mils (.033 in = 33 mils)
1-1/4" flange in 1/100th inches
Track section
3-1/2" member depth in 1/100th inches (inside to inside dimension)

The Right STUFL:

Universal Designator System for Cold-Formed Steel Framing Members

The Logic

The dimensions of the webs and flanges in the designators are used because they convey an intuitive picture of the shape, like a "2 x 4" describes a wood stud. The web depth was put first, instead of the flange width (the "2" in "2 x 4") because with steel the strength-to-weight ratio is usually optimized by increasing web depth instead of flange width, all else being equal.

The web depth and flange width are expressed in 1/100th or 1/1000th inches because fractions are messy, and with such a variety of depths and widths available, rounding to the nearest fractions of an inch is not sufficiently descriptive. Also, the metric system is expected to become universal, in which case the depths and widths would be expressed in millimeters, resulting in the same 3-digit format. Lastly, the logic behind the alpha designators (S for Stud, T for Track, etc.) should be obvious.

Minimum Base Steel Thicknesses

The minimum delivered base steel thickness designators are those used in Prescriptive Standards that are currently in the International Building Code (IBC) and International Residential Code (IRC), and correspond to ICBO's minimum prescribed thicknesses (expressed in mils). These minimum thicknesses are:

Gauge	Inches	Mil
25 gauge	.0179"	18 mil
22 gauge	.0269"	27 mil
20 gauge (Drywall)	.0296"	30 mil
20 gauge (Structural)	.0329"	33 mil
18 gauge	.0428"	43 mil
16 gauge	.0538"	54 mil
14 gauge	.0677"	68 mil
12 gauge	.0966"	97 mil
10 gauge	.1180"	118 mil

Stiffening Lips on C-Sections

The dimension of the stiffening lip by flange width and material thickness are as follows:

Material Thickness	Flange Width	Stiffening Lip
.0179" to .0296"	1-1/4"	3/16"
All thicknesses	1-3/8"	3/8"
All thicknesses	1-5/8"	½"
All thicknesses	2"	5/8"
All thicknesses	2-1/2"	5/8"

The Purpose

With the universal designator system for steel framing members and accepted minimum base steel thicknesses and stiffening lip dimensions, the section properties and load-carrying abilities of any given profile can be calculated and implemented. Regardless of the manufacturer, not only will the designation be identical for a standard steel framing member, but the section properties and load-carrying abilities of that standard member will be uniform throughout the Country. In addition to helping eliminate the confusion in the market stemming from the widely varying properties and loads published by manufacturers making essentially identical shapes, these standards greatly facilitate submittals for plan check, code approvals, prescriptive standards, software development, etc. The intent is to make the products easier to use in existing markets and to accelerate their acceptance in new markets.

Appendix C

Exterior Wall Details

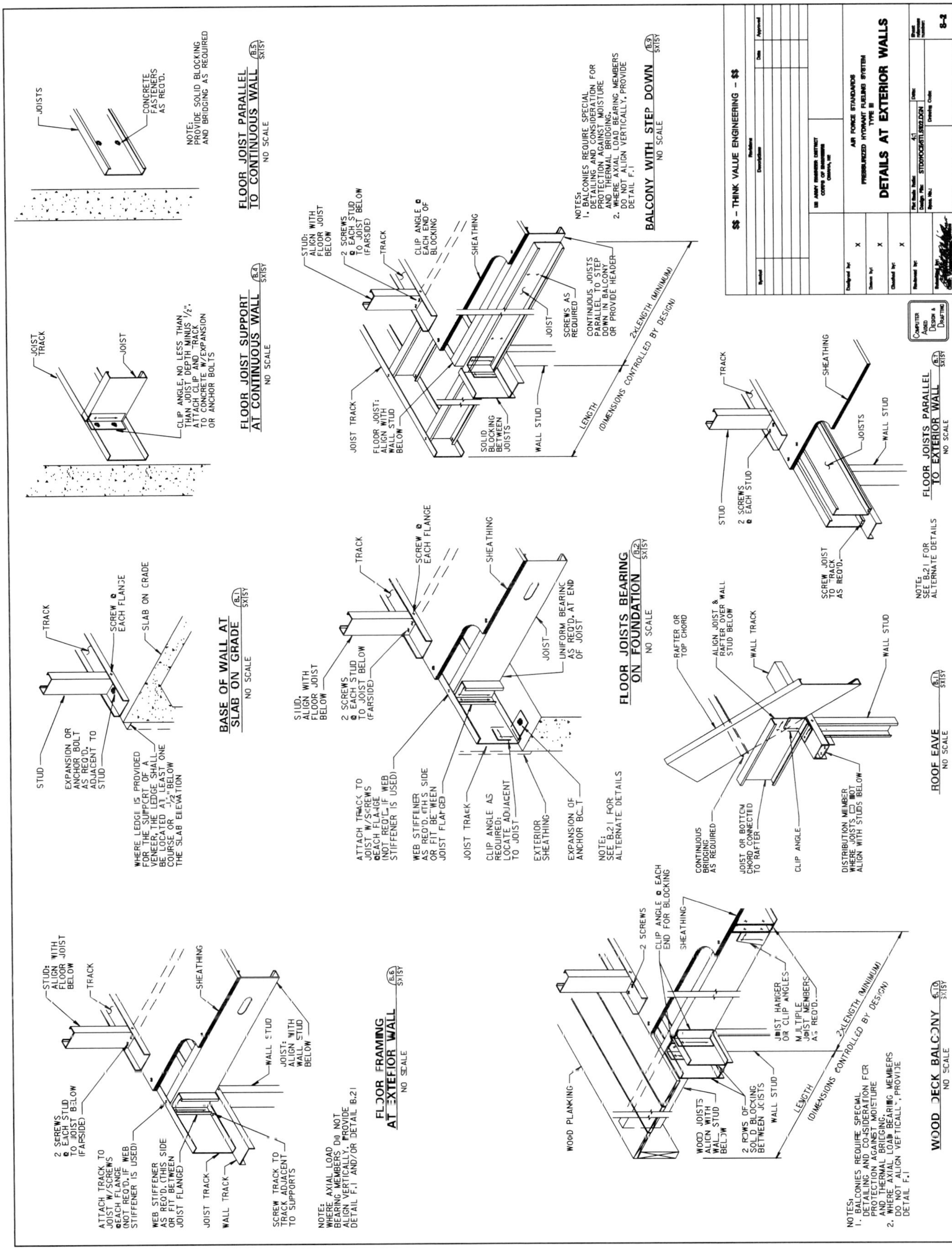
JOISTS
CONCRETE FASTENERS AS REQ'D.
NOTE: PROVIDE SOLID BLOCKING AND BRIDGING AS REQUIRED
FLOOR JOIST PARALLEL TO CONTINUOUS WALL
NO SCALE
JOIST TRACK
JOIST
CLIP ANGLE, NO LESS THAN THAN JOIST DEPTH MINUS 1/2", ATTACH CLIP AND TRACK TO CONCRETE W/EXPANSION OR ANCHOR BOLTS
FLOOR JOIST SUPPORT AT CONTINUOUS WALL
NO SCALE
STUD: ALIGN WITH FLOOR JOIST BELOW
2 SCREWS @ EACH STUD TO JOIST BELOW (FARSIDE)
TRACK
CLIP ANGLE @ EACH END OF BLOCKING
SHEATHING
JOIST TRACK
FLOOR JOIST: ALIGN WITH WALL STUD BELOW
SOLID BLOCKING BETWEEN JOISTS
WALL STUD
JOIST
SCREWS AS REQUIRED
CONTINUOUS JOISTS PARALLEL TO STEP DOWN IN BALCONY OR PROVIDE HEADER
LENGTH
2xLENGTH (MINIMUM)
(DIMENSIONS CONTROLLED BY DESIGN)
NOTES:
1. BALCONIES REQUIRE SPECIAL DETAILING AND CONSIDERATION FOR PROTECTION AGAINST MOISTURE AND THERMAL BRIDGING.
2. WHERE AXIAL LOAD BEARING MEMBERS DO NOT ALIGN VERTICALLY, PROVIDE DETAIL F.1
BALCONY WITH STEP DOWN
NO SCALE
$$ - THINK VALUE ENGINEERING - $$
AIR FORCE STANDARDS
PRESSURIZED HYDRANT FUELING SYSTEM TYPE III
DETAILS AT EXTERIOR WALLS
S-2
STUD
EXPANSION OR ANCHOR BOLT AS REQ'D. ADJACENT TO STUD
TRACK
SCREW @ EACH FLANGE
SLAB ON GRADE
WHERE LEDGE IS PROVIDED FOR THE SUPPORT OF A VENEER, THE LEDGE SHALL BE LOCATED AT LEAST ONE COURSE OR 1/2" BELOW THE SLAB ELEVATION
BASE OF WALL AT SLAB ON GRADE
NO SCALE
STUD: ALIGN WITH JOIST BELOW
2 SCREWS @ EACH STUD TO JOIST BELOW (FARSIDE)
SCREW @ EACH FLANGE
SHEATHING
JOIST
UNIFORM BEARING AS REQ'D. AT END OF JOIST
ATTACH TRACK TO JOIST W/SCREWS @EACH FLANGE (NOT REQ'D. IF WEB STIFFENER IS USED)
WEB STIFFENER AS REQ'D. (THIS SIDE OR FIT BETWEEN JOIST FLANGE)
JOIST TRACK
CLIP ANGLE AS REQUIRED: LOCATE ADJACENT TO JOIST
EXTERIOR SHEATHING
EXPANSION OF ANCHOR BOLT
NOTE: SEE B.21 FOR ALTERNATE DETAILS
FLOOR JOISTS BEARING ON FOUNDATION
NO SCALE
STUD
2 SCREWS @ EACH STUD
SCREW JOIST TO TRACK AS REQ'D.
SHEATHING
JOISTS
WALL STUD
NOTE: SEE B.21 FOR ALTERNATE DETAILS
FLOOR JOISTS PARALLEL TO EXTERIOR WALL
NO SCALE
RAFTER OR TOP CHORD
ALIGN JOIST & RAFTER OVER WALL STUD BELOW
WALL TRACK
WALL STUD
CONTINUOUS BRIDGING AS REQUIRED
JOIST OR BOTTOM CHORD CONNECTED TO RAFTER
CLIP ANGLE
DISTRIBUTION MEMBER WHERE JOISTS DO NOT ALIGN WITH STUDS BELOW
ROOF EAVE
NO SCALE
STUD: ALIGN WITH FLOOR JOIST BELOW
TRACK
SHEATHING
2 SCREWS @ EACH STUD TO JOIST BELOW (FARSIDE)
WALL STUD
JOIST: ALIGN WITH WALL STUD BELOW
ATTACH TRACK TO JOIST W/SCREWS @EACH FLANGE (NOT REQ'D. IF WEB STIFFENER IS USED)
WEB STIFFENER AS REQ'D. (THIS SIDE OR FIT BETWEEN JOIST FLANGE)
JOIST TRACK
WALL TRACK
SCREW TRACK TO TRACK ADJACENT TO SUPPORTS
NOTE: WHERE AXIAL LOAD BEARING MEMBERS DO NOT ALIGN VERTICALLY, PROVIDE DETAIL F.1 AND/OR DETAIL B.21
FLOOR FRAMING AT EXTERIOR WALL
NO SCALE
2 SCREWS
CLIP ANGLE @ EACH END FOR BLOCKING
SHEATHING
JOIST HANGER OR CLIP ANGLES
MULTIPLE JOIST MEMBERS AS REQ'D.
WOOD PLANKING
WOOD JOISTS ALIGN WITH WALL STUD BELOW
2 ROWS OF SOLID BLOCKING BETWEEN JOISTS
WALL STUD
LENGTH
2xLENGTH (MINIMUM)
(DIMENSIONS CONTROLLED BY DESIGN)
NOTES:
1. BALCONIES REQUIRE SPECIAL DETAILING AND CONSIDERATION FOR PROTECTION AGAINST MOISTURE AND THERMAL BRIDGING.
2. WHERE AXIAL LOAD BEARING MEMBERS DO NOT ALIGN VERTICALLY, PROVIDE DETAIL F.1
WOOD DECK BALCONY
NO SCALE

202A03.EPS

Additional Resources and References

Additional Resources

This module is intended to be a thorough resource for task training. The following reference works are suggested for further study. These are optional materials for continued education rather than for task training.

Allied Steel Buildings: www.alliedbuildings.com

Dietrich Metal Framing, Inc.: www.dietrichindustries.com

Steel Framing Alliance: www.steelframing.org

The Steel Network, Inc.: www.steelnetwork.com

Figure Credits

Topaz Publications, Inc., 202F03, 202F06, 202F10, 202F28 (bottom), 202SA01, 202F39

Senco Products, Inc., 202F04

Photo courtesy of Simpson Strong-Tie Company, Inc., 202F05

Steel Framing Alliance, 202F07, 202F13, 202F15, 202F16, 202F18, Table 1, 202F19–202F22, 202F28 (top), 202F29–202F33, 202F35–202F38, 202F43–202F48, 202F50, 202F51, Appendix A and B

The M. K. MORSE Company, 202F08, 202F09

Klein Tools, Inc., 202F11

The Stanley Works, 202F12

The Steel Network, 202F25, 202F26, 202SA03

Radius Track Corporation, 202SA02

U.S. Army Corps of Engineers, Appendix C

NCCER CURRICULA — USER UPDATE

NCCER makes every effort to keep its textbooks up-to-date and free of technical errors. We appreciate your help in this process. If you find an error, a typographical mistake, or an inaccuracy in NCCER's curricula, please fill out this form (or a photocopy), or complete the online form at **www.nccer.org/olf**. Be sure to include the exact module ID number, page number, a detailed description, and your recommended correction. Your input will be brought to the attention of the Authoring Team. Thank you for your assistance.

Instructors – If you have an idea for improving this textbook, or have found that additional materials were necessary to teach this module effectively, please let us know so that we may present your suggestions to the Authoring Team.

NCCER Product Development and Revision

13614 Progress Blvd., Alachua, FL 32615

Email: curriculum@nccer.org

Online: www.nccer.org/olf

❑ Trainee Guide ❑ Lesson Plans ❑ Exam ❑ PowerPoints Other ______________________

Craft / Level: Copyright Date:

Module ID Number / Title:

Section Number(s):

Description:

Recommended Correction:

Your Name:

Address:

Email: Phone:

45203-09

Acoustical Ceilings

45203-09
Acoustical Ceilings

Topics to be presented in this module include:

Overview

Acoustical ceilings are found in most commercial buildings. Unlike fixed ceilings, they provide easy access to wiring, cabling, and air conditioning equipment located in the area between the ceiling and the overhead deck. The ceiling panels are designed to suppress sound transmission. Acoustical ceilings are built by first installing a grid that is suspended from the overhead deck. This grid must be both level and straight; any deviation will make installation difficult and will likely be visible to an observer. Ceiling panels are installed in the grid. There are several types of grids and many types and sizes of ceiling panels. Proper installation of acoustical ceilings is a skill that takes training and practice.

Objectives

When you have completed this module, you will be able to do the following:

1. Establish a level line.
2. Explain the common terms related to sound waves and acoustical ceiling materials.
3. Identify the different types of suspended ceilings.
4. Interpret plans related to ceiling layout.
5. Sketch the ceiling layout for a basic suspended ceiling.
6. Perform a material takeoff for a suspended ceiling.
7. Install selected suspended ceilings.

Trade Terms

Acoustical materials	Diffuser
Acoustics	Fissured
A-weighted decibel (dBA)	Frequency
Ceiling panels	Hertz (Hz)
Ceiling tiles	Striated

Prerequisites

Before you begin this module, it is recommended that you successfully complete *Core Curriculum; Drywall Level One*; and *Drywall Level Two*, Modules 45201-09 and 45202-09.

This course map shows all of the modules in the second level of the *Drywall* curriculum. The suggested training order begins at the bottom and proceeds up. Skill levels increase as you advance on the course map. The local Training Program Sponsor may adjust the training order.

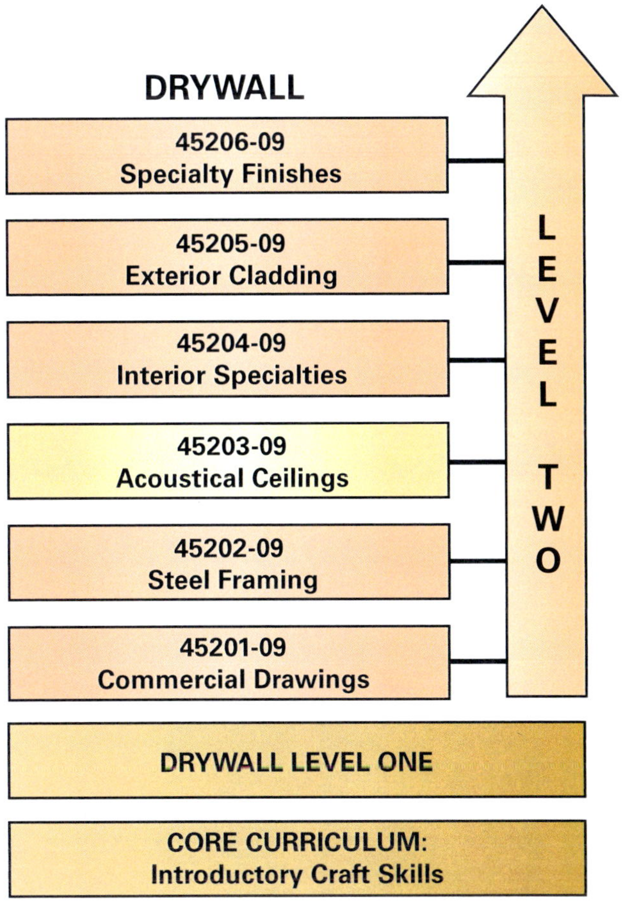

203CMAP.EPS

1.0.0 ♦ INTRODUCTION

Ceilings are not generally considered structural members of a building. In commercial buildings, ceilings are often high to make room for heating ducts, electrical wiring, and plumbing components (*Figure 1*). While this makes for easy maintenance of components, it does little to beautify a work space. Most commercial buildings today use some type of suspension ceiling design to cover unsightly components while still allowing easy access for maintenance.

Suspended acoustical ceilings are widely used in commercial construction, and to some extent, in residential construction. These ceilings serve many purposes. Ceilings are not only used to make a room more attractive, they are designed to help keep outside noise from entering the room and to reduce noise levels occurring within the room. In some cases, ceilings are integrated with the electrical and HVAC (heating, ventilation, and air conditioning) functions to provide for correct lighting and temperature control. Use of attractive **ceiling panels** or **ceiling tiles** helps give a warm, relaxed feeling to a room. Complete ceiling systems offer a wide variety of options, both functional and decorative. The selection of **acoustical materials** depends on the intended use of the room.

203F01.EPS

Figure 1 ♦ Suspended ceiling.

Ceiling systems are selected based on the projected use of an area. The type of ceiling used in a concert hall will be very different from the one used in an office space or in a classroom. As a drywall technician, your job will be to properly install whatever system is shown on the project plans. Improper installation can decrease the beneficial effects of the properties for which the ceiling system was selected. For example, failure to seat ceiling panels fully into their grids can change the **acoustics** of a finished area. Although improperly seating one panel may not change the overall acoustical characteristics of an area, improperly seating multiple panels will.

There is a wide variety of suspended acoustical ceiling systems, with each system being somewhat different from the others. They use the same basic materials, but their appearances are completely different.

This module provides basic information about acoustics and acoustical ceilings. In addition, it covers the standard installation of suspended ceiling systems. The focus is on the following ceiling systems:

- Exposed grid systems
- Metal pan systems
- Direct-hung concealed grid systems
- Integrated ceiling systems
- Luminous ceiling systems
- Suspended drywall furring ceiling systems
- Special ceiling systems

GOING GREEN

Recycling

New ceiling panels may be made from recycled materials, and many old ceiling tiles can be recycled into materials that are used to make new products. Some manufacturers, such as Armstrong, have their own recycling programs. Sometimes the cost of recycling is actually less than the cost of disposal. Recycling is beneficial to the environment and is an important cost-saving measure.

INSIDE TRACK

Look Up

The next time you are out, take a look at the ceilings in the stores, theatres, malls, and other buildings you enter. You will see that the available styles, designs, materials, and color schemes are more numerous than you ever imagined. The one thing they have in common is that most of them are some form of suspended ceiling using panels set into or attached to a framework.

203SA01.EPS

203SA02.EPS

2.0.0 ◆ SOUND WAVE PROPAGATION AND CHARACTERISTICS

Commercial buildings are typically large, sturdy structures that are designed for safety and long years of use rather than for appearance. An easy way for a ceiling to enhance an area is to lower the ceiling, making the area smaller. This makes a large area easier to heat, cool, and light, and it decreases the echo distortion from sound reflection.

One of the most important features of ceiling materials is the effect they have on sound. Sounds travel through air as a series of pressure waves. These sound pressure waves travel outward in all directions. When sound waves strike a wall or ceiling, some of the sound wave energy is absorbed and some is reflected in wave patterns moving in the opposite direction (*Figure 2*). As a result, you will hear both the original sound and its reflection. The sound is also transmitted through the air within the wall or ceiling cavity to the opposite surface, causing the surface to vibrate and transmit the sound to any adjoining rooms.

Sound waves have a **frequency**, or pitch, measured in **hertz (Hz)**. Frequency is the number of vibrations or cycles that occur in the sound wave in one second. The greater the number of cycles per second (Hz), the higher the frequency and the pitch (*Figure 3*).

The intensity of sound refers to its degree of loudness or softness. An **A-weighted decibel (dBA)** is a unit of measure used for establishing and comparing the intensity of sound sources (*Figure 4*). It is used to express the value of all sounds in a range from 0 dBA to 140 dBA and higher.

Table 1 shows some typical examples of noise situations and their relative dBA levels. Note that any sounds greater than 120 dBA can actually produce a physical sensation. Sounds above 130 dBA can cause pain and/or deafness.

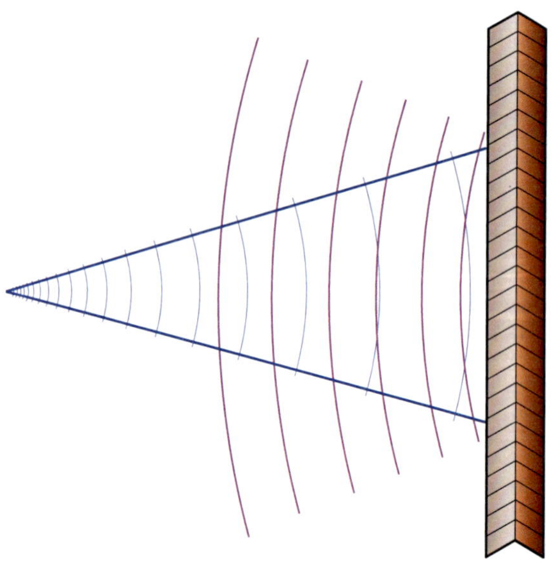

Figure 2 ◆ Sound wave reflection.

3.0.0 ◆ ACOUSTICAL TERMS

Some terms you may encounter and should understand when selecting or working with acoustics and acoustical ceiling materials include the following:

- *Reflection* – The bouncing back of sound waves after striking some obstacle or surface, such as a ceiling or wall.

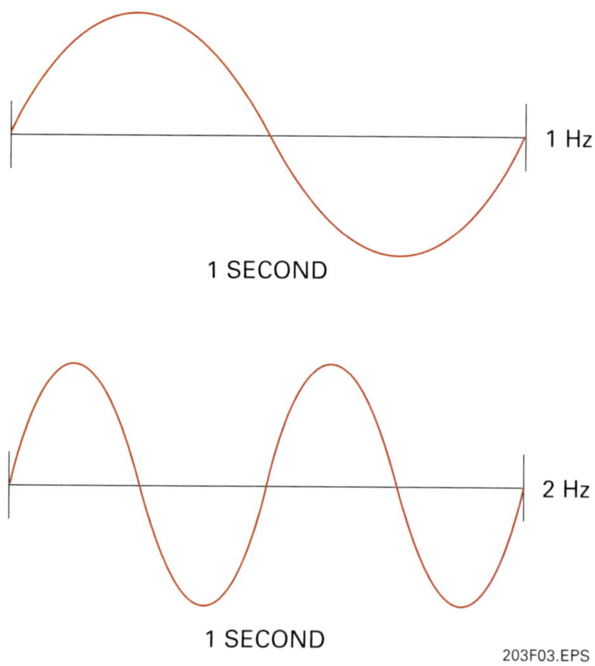

Figure 3 ◆ Frequency.

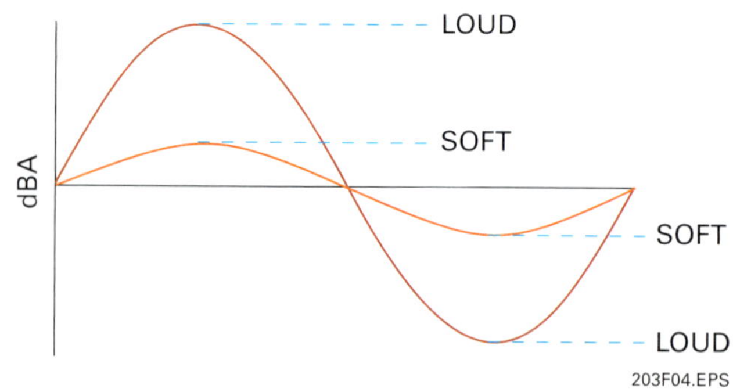

Figure 4 ◆ Sound intensity.

Table 1 Sound Levels of Some Common Noises

Sound Level (dBA)	Intensity Level	Environment	
		Outdoor	Indoor
140	Deafening	Jet aircraft, artillery fire	Gunshot
130	Threshold of pain	—	Loud rock band
120	Threshold of feeling	Elevated train	Portable stereo headset on high setting
110	Extremely loud	Overhead jet aircraft at 1,000'	Loud nightclub
100	Very loud	Chainsaw, motorcycle at 25', auto horn at 10'	—
90	Loud	Lawn mower, noisy city street	Full symphony band, noisy factory
80	Moderately loud	Diesel truck at 50'	Garbage disposal, dishwasher
70	Average	—	Face-to-face conversation, vacuum cleaner, printers and copiers
60	Moderately quiet	Air conditioning condenser, at 15', auto traffic near an interstate highway	Normal conversation, general office
50	Quiet	Large transformer at 50'	—
40	Very quiet	Bird calls	Private office, soft radio music
30	Extremely quiet	Quiet residential neighborhood	Average residence
20	Nearly silent	Rustling leaves	Quiet theater, whisper
10	Just audible	—	Human breathing
0	Threshold of human hearing	—	One's own heartbeat in a silent room

- *Reverberation* – The prolonging of a sound through multiple reflections of the sound as it travels back and forth across a room. These multiple reflections of sound from surfaces such as walls, ceilings, and floors occur so fast that they are usually not heard as distinct repetitions of the sound. However, they can cause a higher noise level in the room than the original sound source.
- *Noise reduction coefficient (NRC)* – Used by manufacturers to compare the noise absorbencies of acoustical products. The higher the number, the better the absorbency. The NRC measures the average percentage of noise a material absorbs at four selected frequencies.
- *Sound transmission loss* – The amount of sound lost as a noise travels through a material. Acoustical ceiling assemblies are rated in terms of sound transmission classification (STC). An STC value of 20 to 25 indicates that even normal speech can be easily understood in an adjoining room. On the other hand, an STC value of 50 to 60 indicates that loud sounds will be heard only faintly or not at all. Acceptable STC ratings range from approximately 39 to 65.
- *Articulation class (AC)* – Rates a ceiling's ability to achieve normal privacy in open office spaces by absorbing the noise reflected at an angle off of the ceiling into adjacent areas (cubicles). Per the *ASTM E1110* and *E1111* standards, the generally accepted AC ratings for normal privacy in open plan offices are a minimum of 170, with 190 to 210 preferred.
- *Ceiling attenuation class (CAC)* – Rates a ceiling's efficiency as a barrier to airborne sound transmission between adjacent work areas, where sound can penetrate plenum spaces and travel to other spaces. Sound is capable of traveling to rooms far from the source of the original sound. CAC is stated as a minimum value. Per *ASTM Standard E1264*, CAC minimum 25 is acceptable in an open plan office, while a rating of CAC minimum 35 to 40 is preferred for closed offices.
- *Absorption* – Describes the soaking up or taking in of sound wave energy by a material's surface, as opposed to the sound waves being reflected or bounced.

4.0.0 ◆ READING PLANS AND SPECIFICATIONS

Ceilings in commercial construction are designed to look attractive, cover unsightly mechanical and structural components, and enhance the surroundings. Part of your job as a drywall technician will be to read and understand the project plans and specifications. These skills are essential as you work with other specialty crafts in the installation of devices such as lighting fixtures, sprinkler systems, and air **diffusers** that protrude through ceilings.

To avoid mistakes and omissions when installing ceilings, an organized and systematic approach should be used when reading the related plans. The following is a general procedure for reading plans:

Step 1 Check the room schedule on the plans.
- Identify the type of ceiling material to be used.
- Locate the protrusions into the ceiling.

Step 2 Locate the room on the floor plan.
- Find the room dimensions. If none are found, locate the drawing scale.
- Using the given scale, determine the dimensions of the room.

> **NOTE**
> When no dimensions are given on a diagram, using the scale to calculate room dimensions will provide a rough idea of the room size. To obtain accurate dimensions, check with your supervisor or measure the room.

Step 3 Check to see if there is a reflected ceiling plan. If no reflected ceiling plan is found, check to see if a shop drawing is included.

Step 4 Be sure the plans are the final revised set.
- Plans are often revised several times before the ceiling is ready to be installed. To be sure the plans are the final revised set, check the date of issuance against the work order to see if they are the same.
- If work that has already been done is not reflected on the plans, chances are the plans are not the final revised set. This could have a significant impact on the ceiling installation.

Step 5 Read the project specifications.
- Be sure the ceiling to be installed is the same as that listed in the specifications.
- If the job conditions do not agree with what is shown in the specifications and/or plans, call your department superintendent and ask for instructions on how to proceed.

Step 6 Check the mechanical and electrical plans prior to the layout of the ceiling.
- On the mechanical plans, locate the air diffusers (HVAC air supply outlets), return grilles, ducts, and sprinkler heads.
- On the electrical plans, locate the light fixtures, fans, and other ceiling protrusions.

On large construction jobs, there may be one or more reflected ceiling plans (RCP). *Figure 5* is an example of a typical RCP. It shows the ceiling as viewed from above, looking down toward the floor. RCPs show all ceiling fixtures—lights, ceiling fans, air diffusers and returns, sprinklers—as well as the ceiling tiles. RPCs also show details of how a ceiling will be constructed. A typical plan includes the following:

- Elevation of the ceiling above floor level
- Location of the center (starting) line
- Size of the borders
- Position of the light fixtures
- Location of the air diffusers

As a rule, reflected ceiling plans also show all items that penetrate the ceiling, including return grilles, diffusers, sprinkler heads, recessed lights, and recessed speakers for sound systems.

In addition to the project drawing set, you may need to refer to shop drawings provided by the manufacturer of the ceiling system. These drawings show more details about the system components and installation than are shown in the project drawings. When you find a difference between the project drawings and the manufacturer's instructions, be sure to clarify the discrepancy with your supervisor before installing the ceiling.

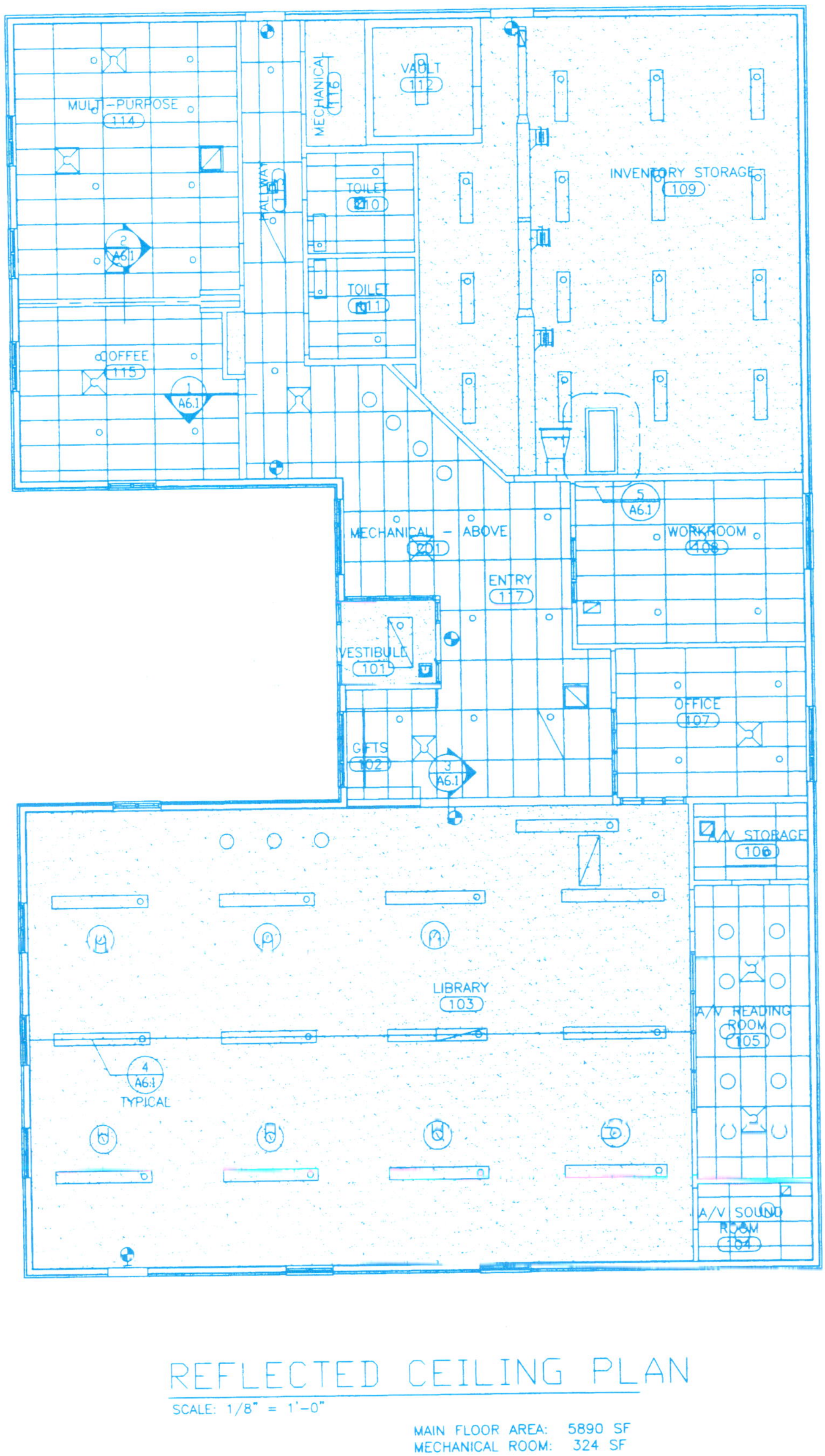

203F05.EPS

Figure 5 ◆ Reflected ceiling plan.

INSIDE TRACK

Plenum Ceilings

The systems that provide heating and cooling for most commercial buildings are forced-air systems. Blower fans are used to circulate the air. The blower draws air from the space to be conditioned and then forces the air over a heat exchanger, which cools or heats the air. In a cooling system, for example, the air is forced over an evaporator coil that has very cold refrigerant flowing through it. The heat in the air is transferred to the refrigerant, so the air that comes out the other side of the evaporator coil is cold. In homes, the air is delivered to the conditioned space and returned to the air conditioning/heating system through ductwork that is usually made of sheet metal. In commercial buildings with suspended ceilings, the space between the ceiling and the overhead decking is often used as the return air plenum. It is often called an open plenum. (A plenum is a sealed chamber at the inlet or outlet of an air handler.) This approach saves money by eliminating about half the cost of materials and labor associated with ductwork.

One factor to keep in mind is that anything in the plenum space (electrical or telecommunications cable, for example) must be specifically rated for plenum use to meet fire ratings. Plastic sheathing used on standard cables gives off toxic fumes when burned. Plenum-rated cable uses non-toxic sheathing.

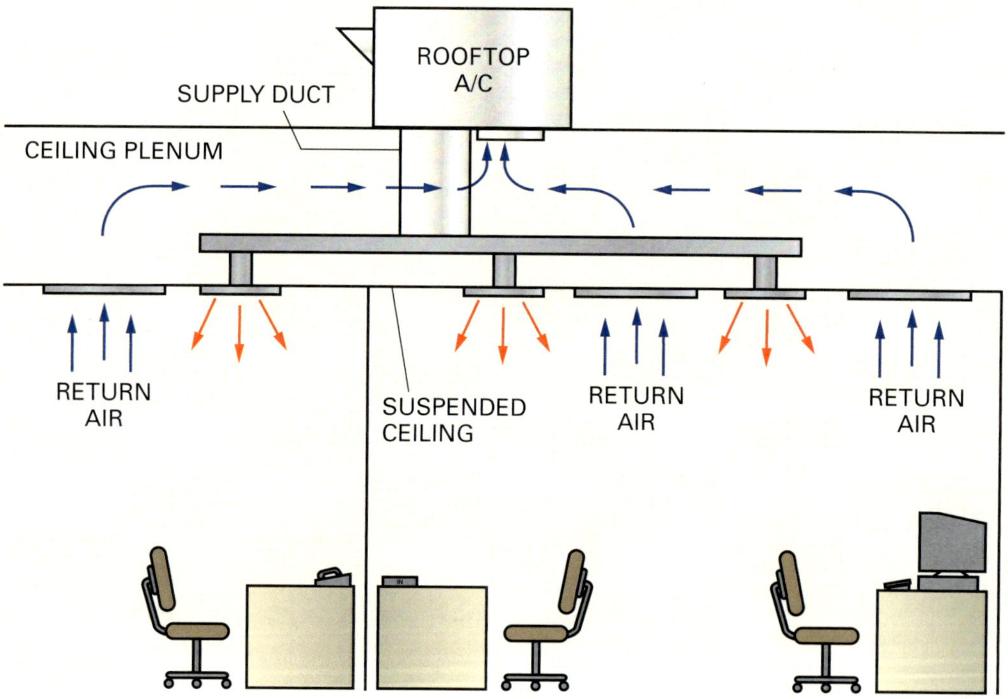

203SA03.EPS

5.0.0 ◆ CEILING LEVELING EQUIPMENT

For the installation of suspended ceilings, the most popular leveling device is the laser. Many types of laser leveling devices can be used to aid in the installation of suspended ceilings. To use a laser to level a ceiling, set up and operate the laser level according to the manufacturer's instructions for the laser being used. Generally, the procedure involves mounting the laser either on a wall/ceiling mount (*Figure 6*) or on a tall tripod. The laser beam is rotated either at the finished ceiling height or at a reference point. A special target is snapped to a grid, then the grid is moved up or down until the laser beam crosses the target's offset mark. The grid is then secured in place.

Figure 6 ◆ Laser instrument wall/ceiling mount.

WARNING!

OSHA has published *Standard 1926.54* for the safe use of lasers. Some guidelines are as follows:

- Avoid direct eye exposure to the laser beam, even when using a low-power laser.
- Only qualified and trained personnel should operate laser equipment.
- Place a standard laser warning sign conspicuously at major approaches to the instrument use area.
- Always turn off the laser when transmission of the beam is not required.

6.0.0 ◆ CEILING PANELS AND TILES

Ceiling panels and tiles range in size from 12" × 12" up to 60" × 60". Various colors and designs are available. Most are fabricated from mineral and glass fibers. Depending on their design and purpose, mineral fiber panels and tiles are made with painted, plastic, aluminum, ceramic, and mineral faces. Glass fiber panels and tiles are made with painted, film, glass cloth, and molded faces.

There are basically three types of panel/ceiling grid interfaces: lay-in, concealed tee, and profiled edge. Lay-in panels are widely used and are generally the most cost-effective style. Concealed-tee panels are butt-jointed panels that provide a monolithic ceiling design with no visible support system. Profiled-edge panels are made in a wide selection of edge designs from soft-edged chamfered or curved panels to highly articulated edges. Beveled and angular tegular (reveal) edged panels provide a three-dimensional look in a suspended ceiling.

Many interior spaces have specific requirements for ceiling panel materials and characteristics. These can include sound control, fire resistance, thermal insulation, moisture resistance, and light reflectance. Other considerations include maintenance, appearance, and cost.

Noise is often considered to be the greatest distraction in the workplace. Choosing the right ceiling material can help reduce office noise. High- performance acoustical panels are used to help reduce background noise in open plan and closed types of offices. The following three factors contribute to noise distractions in a workplace:

- *General office noise* – The ability of a ceiling material to absorb general office noise is measured using a value known as the noise reduction coefficient (NRC). This value is used by panel manufacturers to compare the noise absorbency of their ceiling panel products. The higher the NRC number, the better the absorbency. The NRC measures the average percentage of noise a panel absorbs at various frequencies.
- *Reflected conversational noise that angles off ceilings into adjacent cubicles* – The ability of a ceiling material to absorb reflected conversational noise is measured using a value known as the articulation class (AC). This value rates a ceiling's ability to achieve normal privacy in open office spaces by absorbing noise reflected at an angle off the ceiling into adjacent areas (cubicles). Per *ASTM E1110* and *E1111*, the generally accepted AC ratings for normal privacy in open plan offices are a minimum of 170, with 190 to 210 preferred.
- *Sound transmission through cubicles, partitions, walls, and ceilings* – The ability of a ceiling material to absorb sound transmission is measured using a value known as the ceiling attenuation class (CAC). This rates a ceiling panel's efficiency as a barrier to airborne sound transmission between adjacent work areas, where sound can penetrate plenum spaces and carry to other spaces. The CAC is stated as a minimum value. Per *ASTM E1264*, CAC minimum 25 is acceptable in an open plan office, while a rating of CAC minimum 35 to 40 is preferred for closed offices.

Local building codes require fire-safe construction for many building applications. Fire-resistant ceiling panels and support systems are specially made of materials that provide increased resistance against flame spread, smoke generation, and/or structural failure in the event of a fire. Two ratings based on ASTM, ANSI, and NFPA standards are used to evaluate fire-resistant panels: the flame spread rating of the material (*ASTM E84*) and the fire resistance rating of the ceiling assembly (*ANSI/UL 263, ASTM E119,* and *NFPA 251*).

The flame-spread rating of a ceiling material is the relative rate at which a flame will spread over the surface of the material. This rate is compared against a rating of 0 (highest rating, meaning that fire spreads slowly) for asbestos-cement board and a rating of 100 for red oak. Class A ceilings have flame-spread ratings of 25 or less, the required standard for most commercial applications. The fire-resistance rating of a ceiling assembly represents the degree to which the entire assembly, not the individual components, will withstand fire and high temperatures (measured in hours). Specifically, it is an assembly's ability to prevent the spread of fire between spaces while retaining structural integrity.

Most ceiling tiles provide little thermal insulation between the space above the suspended ceiling and the work area below. When the building design or local codes require insulation above ceiling panels, special care must be taken to avoid placing too much weight on the ceiling system. When insulation is required, fiberglass ceiling panels are recommended. Most manufacturers discourage the use of insulation on mineral ceiling panels because the panels may sag as the humidity increases. If insulation is necessary on mineral ceiling panels, it should be limited to R-19 roll insulation, put in with the vapor barrier facing down, and installed perpendicular to the cross tees so they are supporting most of the weight. Insulation should not be installed above fire-resistant ceiling panels.

High humidity resistance panels are panels that have superior resistance to sagging caused by high humidity conditions. Many also inhibit the growth of mold or mildew that may appear in high humidity. Sagging not only diminishes the attractiveness of a ceiling, but it also causes ceilings to chip and soil more easily, reducing the high light reflectivity of the panels. High humidity resistance panels are typically installed in highly humid climates or in areas such as kitchens, locker rooms, shower areas, and indoor pools. Other common locations for panels include buildings where the HVAC systems may be shut down for extended periods, or those in which the ceiling might be installed early in the construction process before the building is fully enclosed.

Ceiling panels with a high light reflectance (LR) value of 0.83 or greater per *ASTM E1477* help increase effective lighting levels and reduce light fixture costs and energy consumption, especially with indirect lighting systems. Their use also helps to reduce eyestrain. These panels typically have soil-resistant surfaces that stay cleaner longer than standard ceilings, resulting in a much lower loss of light reflectance over time.

Detailed information about the various panel characteristics described in this section can normally be found in ceiling manufacturers' product catalogs and literature. When selecting panels for a particular application, it is best to follow the manufacturer's recommendations.

7.0.0 ♦ EXPOSED GRID SYSTEMS

The following sections describe the components of exposed grid ceiling systems, and the tools, materials, and general procedures for installing them. Exposed grid ceiling systems are also called direct-hung systems.

INSIDE TRACK

Fire-Rated Applications

Suspended ceiling systems can be used in fire-rated applications. Factors such as resistance to fire and flame spread for the panels, as well as for the suspension system, are taken into account. However, the ceiling materials alone do not determine the fire rating. The materials and construction methods used in the entire system, including the floor/ceiling or ceiling/roof assembly, all enter into the fire rating determination. If a suspended ceiling is to be part of a fire-rated system, you must consult the manufacturer's product literature to determine the ceiling panel and grid that must be used to meet the rating.

High-Durability Ceilings

INSIDE TRACK

Highly durable ceiling panels provide for long life and easy maintenance in places where ceilings are subjected to improper use, vandalism, or frequent removal for plenum access. They are used in applications where resistance to impact, scratches, and soil are major considerations. Ceilings in areas such as school corridors or gymnasiums need to withstand abuse, including surface impact. In areas where lay-in ceiling panels frequently need to be removed for plenum access, scratch-resistant panels are highly desirable. Otherwise, the panel surface can be scratched, scuffed, or chipped as it is slid across the metal suspension system components. Ceilings installed in laboratories, clean rooms, and food preparation areas are normally required to meet special standards to ensure that they can withstand repeated cleaning.

NOTE

The general procedures given in this module are provided as examples only. Because of the differences in components made by different manufacturers, it is important to always follow the manufacturer's installation instructions for the specific system being used.

7.1.0 Grid System Components

In an exposed grid suspended ceiling, a light metal grid is hung by wires that are attached to the original ceiling. Panels that are usually 2' × 2' or 2' × 4' are then placed in the frames of the metal grid. Exposed grid systems are constructed using the components and materials described as follows, and as shown in *Figure 7*.

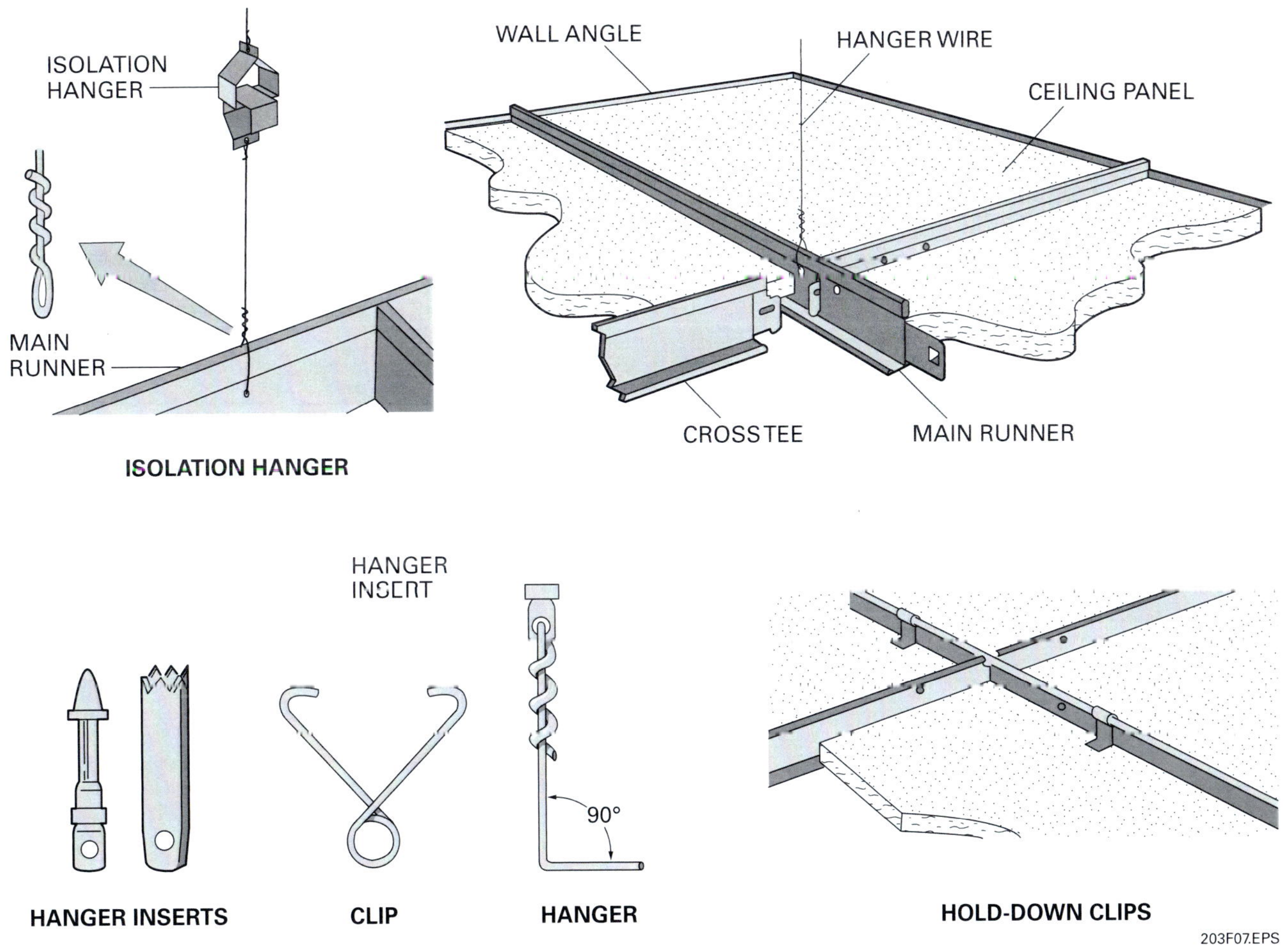

Figure 7 ◆ Exposed grid system components.

"Seeing" Through the Ceiling

INSIDE TRACK

It is a good idea to keep a package of colored thumbtacks in your toolbox when installing a suspended ceiling. When you know that other trades may have to get back in to complete their work, you can insert different-colored thumbtacks in ceiling panels to mark the locations of electrical and mechanical services. The markings will allow the other trades to locate their components without having to raise a lot of panels, and will help minimize finger smudges and damage to the panels.

- *Main runners* – The primary support members of the grid system for all types of suspended ceiling systems. They are 12 feet in length and are usually constructed in the form of an inverted T. When it is necessary to lengthen the main runners, they are usually spliced together using extension inserts; however, the method of splicing may vary with the type of system being used.
- *Cross runners (cross ties or cross tees)* – Are inserted into the main runners at right angles and spaced an equal distance from each other, forming a complete grid system. Cross runners are held in place by either clips or automatic locking devices. They are usually 4 feet or 2 feet in length and are normally constructed in the form of an inverted T. Note that 2-foot cross runners are only required for use with 2' × 2' ceiling panels.
- *Wall angles* – Installed on the walls to support the exposed grid system at the outer edges.
- *Ceiling panels* – Panels that are laid in place between the main runners and cross ties to provide an acoustical treatment. The acoustical panels used in suspended ceilings prevent sound reflection and reverberation by absorbing sound waves. These panels are typically designed with numerous tiny sound traps consisting of drilled or punched holes or fissures, or a combination of both. When sound strikes the panel, it is trapped in the holes or fissures. A wide variety of ceiling panel designs, patterns, colors, facings, and sizes is available, allowing most environmental and aesthetic demands to be met. Panels are typically made of glass or mineral fiber. Generally, glass fiber panels have a higher sound absorbency than mineral fiber panels. Panel facings are typically embossed vinyl in a range of patterns such as **fissured**, pebbled, or **striated**. Due to variations in manufacturers' standards, the specific ceiling panels used must be compatible with the ceiling suspension system.
- *Hanger inserts and clips* – Many types of fastening devices are used to attach the grid system hangers or wires to the building ceiling located above the suspended ceiling. Screw eyes are commonly used with wood framing and some metal decks. Powder-actuated fasteners are also used to fasten into reinforced concrete. Clips are used where beams are available. Clips are typically installed over the beam flanges, then the hanger wires are inserted through the loops in the clips and are secured.
- *Hangers* – The devices attached to the hanger inserts, which are used to support the suspended ceiling's main runners. The hangers can be made of No. 12 wire or heavier rod stock. Ceiling isolation hangers are also available to isolate the ceiling from the noise traveling through the building structure.
- *Hold-down clips* – Used in some systems to hold the ceiling panels in place.
- *Nails, screws, rawls, tappets, and molly bolts* – Used to secure the wall angle to the wall. The specific fastener used depends on the wall construction and material.

7.2.0 Installation Procedure

Doing overhead work often requires the use of scaffolding or ladders to reach the work area. Working even a few feet above ground adds an element of danger to the job. It is important to always place safety first. Use equipment that is the correct height for the job, and be sure that all equipment is in good working order. Any defective items must be tagged right away so they will not be used by mistake. Scaffolding must be inspected by a competent person before it is used on each shift, whenever it is moved, and whenever it is changed in any way. Some scaffolding must be designed by an engineer, so know the practices at your work site. If you are unsure of the site requirements, ask your supervisor or the foreman before you begin work.

Inside Track: What's in a Name?

The terms *ceiling panel* and *ceiling tile* have specific meanings in the trade. Ceiling panels are typically any lay-in acoustical board that is designed for use with an exposed mounting system. They do not have finished edges or precise dimensional tolerances because the exposed grid system support members provide the trimout. Ceiling tiles are acoustical ceiling boards, usually 12" × 12" or 12" × 24", which are nailed, cemented, or suspended by a concealed grid system. The edges are often kerfed and cut back.

When working aboveground, always wear the appropriate personal protective equipment. Maintain control of materials and tools, and stay alert for personnel working on the ground below you. Resist the urge to over-reach for your equipment. If working on the ground when overhead work is being done, stay alert for falling debris and tools.

To install an exposed grid suspended ceiling system, follow these guidelines:

CAUTION

Wear gloves when cutting metal ceiling components.

Step 1 Check the room number and location.

- Make sure that the correct ceiling is going into the correct room.
- Refer to the plans, reflected ceiling plan, or shop drawing to determine the correct height for the ceiling.
- Check the layout to make sure there will be no border tiles under 6 inches. Check with your supervisor and adjust as needed.
- Make sure that the designated ceiling height will work with any installed equipment.
- Check the electrical prints to be sure that the lights will be placed as indicated on the reflected ceiling plan or shop drawing.
- Check the specification sheets or shop orders for special instructions, to ensure that the proper hangers and fasteners will be used.

NOTE

Scaffolds must be used and assembled in accordance with all local, state, and federal/ OSHA regulations. OSHA requires that the building, moving, or dismantling of all scaffolding must be supervised by a competent person who has the training, knowledge, and experience to identify hazards on the job site and the authority to eliminate them. Mobile scaffolds should only be used on level, smooth surfaces that are free of obstructions and openings. OSHA regulations also require that mobile scaffold casters have positive locking devices to hold the scaffold in place. When moving a mobile scaffold, apply the moving force as close to the scaffold base as possible to avoid tipping it over. Never move a scaffold when someone is on it.

Step 2 Install the scaffolding.

- Set the scaffolding at the correct height to permit the driving of inserts or other fasteners into the overhead ceiling and to allow the connection of hangers.
- Set the lower portion of the scaffold to permit installation of the grid members and ceiling panels.

Step 3 Establish the correct and level ceiling height using a laser.

7.2.1 Installation of the Wall Angle

When installing an exposed grid ceiling system, the first task is to install the wall angle. To install the wall angle, do the following:

- Make sure that the top of the wall angle is even with the chalkline at all points.
- Nail or screw the wall angle to the wall. The wall angle should fit securely to prevent sound leaks. Nails or screws must be installed per the manufacturer's instructions.
- Miter the wall angle to fit at the corners or use corner caps to cover the joints.

CAUTION

Wall angle is sharp and can cause cuts. Wear gloves when cutting metal ceiling components.

Indoor Swimming Pools

INSIDE TRACK

All-aluminum grid systems are not recommended for use above indoor swimming pools because chlorine causes aluminum to corrode.

7.2.2 Room Layout

In some cases, reflected ceiling plans or shop drawings may indicate a center line and specify border width. In these instances, follow the suggested layout. If you have neither to follow, you will need to establish the center line and plan the layout of the ceiling so that the border panels adjacent to the facing wall are the same width and are not less than one half the width of a full panel.

One way to determine the width of the border panels is to convert the wall measurement from feet to inches, then divide this amount by the width of the ceiling panels. For example, assume the room measures 42'-6" × 30'-6". What will be the width of the border panels running parallel to the two 42'-6" walls if 2' × 2' panels are used? To find the answer, proceed as follows:

Step 1 Take the measurement of the short wall (30'-6"), and convert feet to inches.

30' × 12" = 360" + 6" = 366"

Step 2 Divide that amount by 24" (the width of a tile).

366" ÷ 24" = 15 tiles
with a remainder of 6"

Step 3 If the division does not result in a whole number, add the width (in inches) of a full board to the remainder (in this case, it is 6").

6" + 24" = 30"

Step 4 Divide this by 2. The result is 15. This would be the width of a border panel on each side. When adding the width of a panel to the remainder in Step 3, you must delete one full panel from the total. This means that there would be 14 full panels (15 – 1) and two 15-inch border panels (*Figure 8*), so you will need 16 panels per row.

14 × 2' + 30" = 30'-6"

Step 5 Determine the width of the border panels for the other two remaining walls in the same manner.

7.2.3 Hanger Inserts

Once all border unit measurements have been made, it is necessary to install ceiling clips, also known as hanger clips, to secure the hangers. Pri-

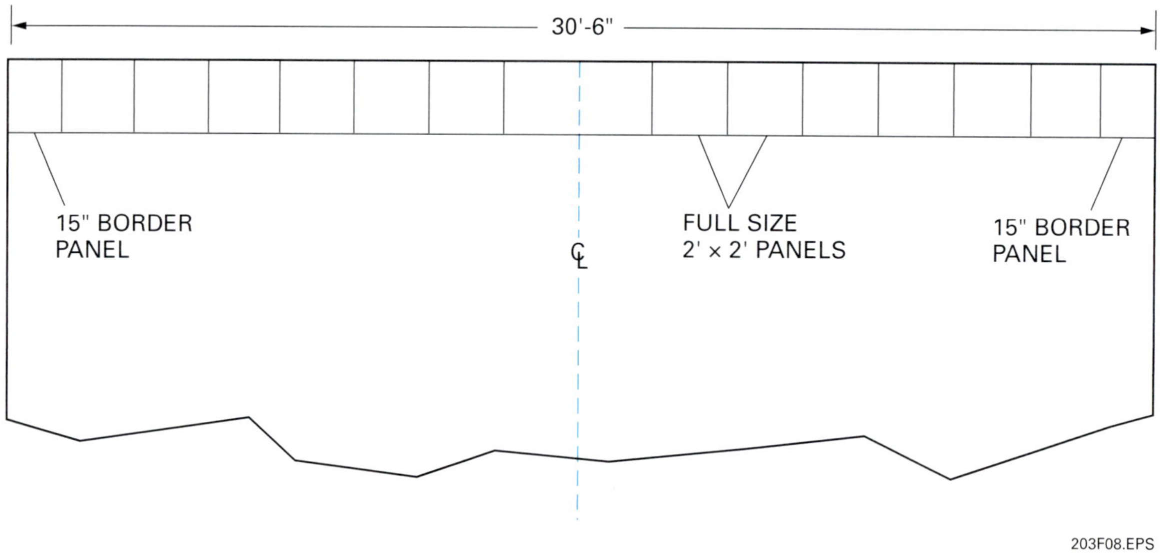

Figure 8 ♦ Fitting the panels to the room.

or to installing the ceiling clips, you must locate the position of the main runners. Line A in *Figure 9* is the main runner.

NOTE

Check the lighting layout for the main T locations. Avoid locating a main T splice next to a fixture.

Step 1 After consulting the reflected ceiling plan, establish the layout of the main T. Starting 6 inches from the wall, install hanger clips/wires every 4 feet on center. Be sure to install one within 6 inches of the opposite wall.

NOTE

Mark the location of the main Ts on the first row floor. This will provide a reference point for installation.

Step 2 Install the next row 4 feet from the first. Proceed until the installation is complete.

WARNING!

Powder-actuated fastening tools are to be used only by trained operators in accordance with the operator's manual. Operators must take precautions to protect both themselves and others when using powder-actuated tools, as follows:

- Operate the tool as directed by the manufacturer's instructions, and use it only for the fastening jobs for which it was designed.
- To prevent injury or death, make sure that the drive pin cannot penetrate completely through or exit the material into which it is being driven.
- To prevent a ricochet hazard, make sure the recommended shield is in place on the nose of the tool.

Use pretied wire wherever possible. If not using pretied wire, as you install each ceiling clip and wire, allow enough wire to ensure proper tying to the ceiling clip and the main runner. These two ties will require from 10 to 24 inches of wire. More hanger wire may be required if the ceiling is hung in part from structural members along with

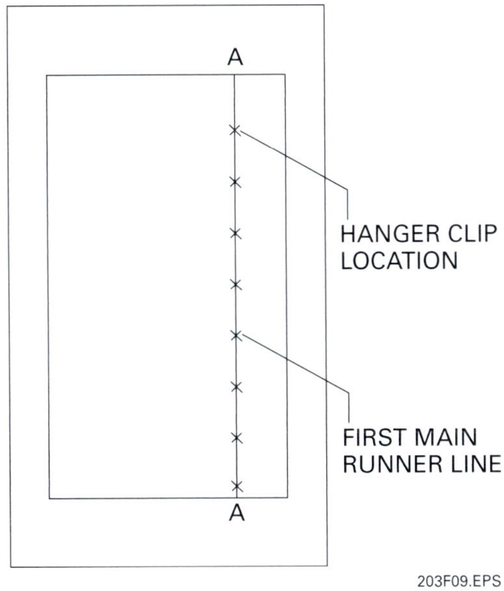

Figure 9 ♦ Locating the position of the first main runner.

the ceiling clips. All hanger wire is precut and can be obtained in various lengths and gauges. Refer to the specifications for the gauge of hanger wire to use. There are cases in which there may be cast-in-place inserts or hanger wires. In such cases, the insertion of ceiling clips and installation of wire will not be necessary.

NOTE

If an obstruction such as duct prevents the installation of hanger wire, trapeze hangers may need to be installed. Avoid running wires at angles. This creates tension that can pull the grid out of square.

7.2.4 Installing the Main Runners

Since the main runners are by far the most difficult line, they are hung first. Install the main runners as follows:

Step 1 Measure and cut the main runner so that the cross runner will be at the proper distance for the border panel.

NOTE

At least three tight turns should be made when twisting the wire. The wire should also be tight to the runner.

Step 2 Bend the hanger wire and suspend the first length of main runner from the first row of hangers. It is important that the first few hanger wires be straight (perpendicular to the main runner). See *Figure 10*. If they are not, use a Whitney punch and make new holes.

NOTE

As indicated previously, in most exposed grid systems, the main runners come prepunched with holes. The upper edges of the holes generally measure 1½ inches from the bottom of the runner. The laser will have to be adjusted prior to bending the hanger wire.

Step 3 Continue to install the balance of the first row of main runners.

Step 4 After installing the last full length of the main runner, measure, cut, and install an end piece from a full length of main runner to complete the first run.

Hold-Down Clips

INSIDE TRACK

Some ceiling panels require hold-down clips to secure the ceiling panels to the grid. For example, clips are used with lightweight panels to prevent them from reacting to drafts. One manufacturer specifies that clips are to be used if the panels weigh less than one pound. Hold-down clips are not necessarily required for ceilings used in fire-rated applications. Check the manufacturer's instructions.

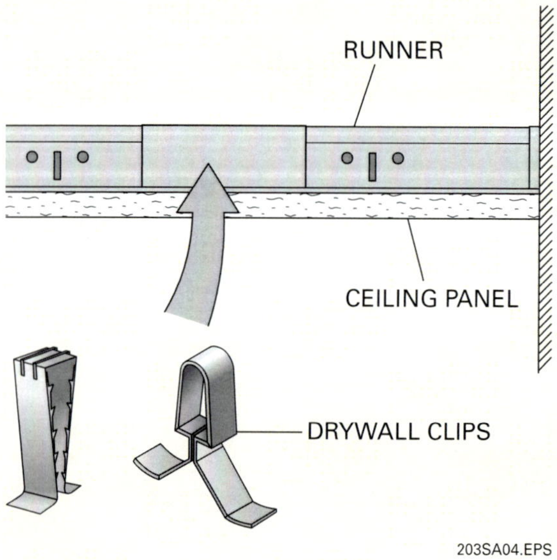

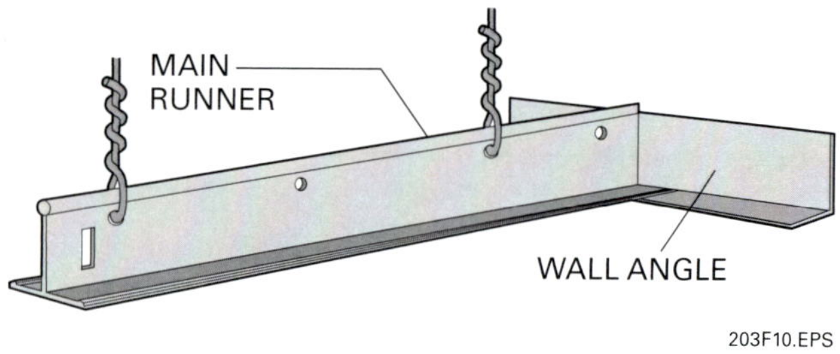

Figure 10 ◆ Inserting hanger wires in a main runner.

Step 5 After each end of the main runner is resting on the wall angle, insert all hanger wires. Do not twist to allow room for adjustment when the ceiling is squared/aligned.

7.2.5 Installing the Cross Runners (Cross Tees)

When installing an exposed grid ceiling system, the last task is to install the cross runners. Use the following procedure to install the cross runners:

Step 1 Measure the distance from the main runner to the wall angle. This width was determined earlier when completing the room layout.

Step 2 Using the snips, cut the cross runner to the correct border width. Be sure to cut and save the correct end or it will not slide into the main runner.

Step 3 Insert the factory end into the main runner and let the other end rest on the wall angle (*Figure 11*).

Step 4 When the main runner and the cross runner are perpendicular to each other, lock the cross runner in place. Install the remaining cross runners between the main runners.

Step 5 To stabilize the grid system, install pop rivets at every other cross runner.

Step 6 Make sure the grid is square.

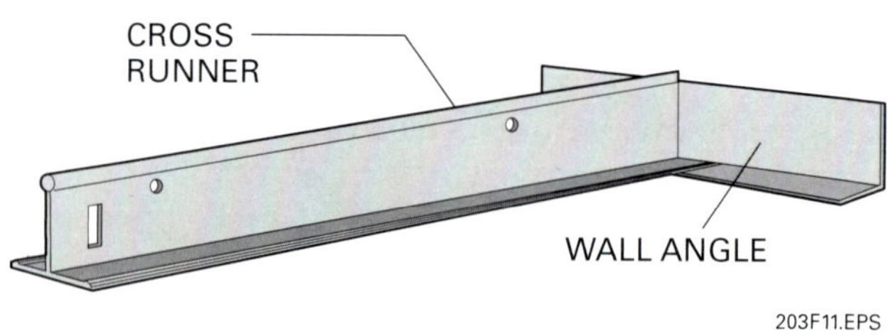

Figure 11 ◆ End of cross runner resting on wall angle.

NOTE

Some grid systems use automatic locking devices to secure the cross runners into the main runners. Other grid systems use clips. Once all cross runners have been installed into the main runners, the grid system is ready to receive the ceiling panel or tile.

7.2.6 Cutting and Placing Ceiling Panels

Prior to installing the ceiling panels, wash your hands before you begin and then periodically during the job to keep the panels clean. It is a good practice to wear clean white gloves when handling ceiling panels. Start by cutting the border panels and inserting them into position. After the border panels have been installed, install the full panels. As the full panels are being placed, install hold-down clips, if required. General installation guidelines for ceiling panels include the following:

- Install the ceiling panels in place according to the job progression. Do not skip panels or randomly scatter them in the grid system.
- Take care when installing panels to prevent damage to or marring of the surface.
- Always handle panels at the edges, and keep your thumbs from touching the finished side of the panels. If gloves are not worn, frequently apply cornstarch, powder, or white chalk to your hands.
- If the ceiling panel has a deeply textured pattern, insert the panels into the grid system so that the pattern flows in the same direction. Such matching of ceiling panels adds beauty to the finished ceiling. Some tiles include a directional arrow.

INSIDE TRACK **Insulation**

It is usually advised that insulation not be placed over mineral ceiling panels. This is because the additional weight could cause the panels to sag. If the use of insulation cannot be avoided because of occupancy codes, limit the insulation to R-19 (0.26 pounds per square foot). Use only roll insulation, and lay it perpendicular to the cross tees. This way the grid supports the weight of the insulation. If batts are required by code, 24" × 24" ceiling panels should be used. Check codes carefully before applying insulation in a fire-rated system.

After completing the ceiling, clean up the work area as follows:

Step 1 Dismantle the scaffold.

Step 2 Pick up all tools and equipment.

Step 3 Secure all equipment in a safe place.

Step 4 Pick up all trash.

Step 5 Sweep up any dust or debris.

8.0.0 ◆ SUSPENDED DRYWALL CEILING SYSTEM

There are two types of suspended drywall ceiling systems: the drywall furring system and the drywall grid system. Suspended drywall systems can be used to create a variety of ceiling effects, as described in the following sections.

8.1.0 Drywall Furring Ceiling System

A suspended drywall furring system is used when it is desirable or specified to have a drywall finish or drywall backing for an acoustical tile ceiling. The installation of a suspended furring system requires the use of special carrying channels.

8.1.1 Installing Carrying Channels

Carrying channels are used for drywall furring systems to compensate for the additional weight of the systems. To install carrying channels, proceed as follows:

Step 1 Snap chalklines on all walls or use a laser level to establish the reference line.

Step 2 Install the wall angle, if used.

Step 3 Hang dry lines. From the ends of two facing walls, measure 4 feet along the chalkline reference line. Secure the dry lines to the walls at these two points.

Step 4 Fasten the hanger inserts into the ceiling structure above. Using the dry line as a guide, install a row of hanger inserts 4 feet on center, or as specified by the manufacturer, into the ceiling over the dry line.

Step 5 Install the hangers. Use No. 8 gauge wire or as specified. Hang the wires from the hanger inserts, then twist and secure them. Allow enough wire to tie the hanger to the insert and the carrying channel. In general, 24 to 28 inches of wire is adequate.

Step 6 Mark the places where the dry line or laser target touches the hanger wires. Mark each hanger wire at the spot where it is to be bent. Bend the wire to a 90-degree angle.

Step 7 When all the hanger wires in one row have been marked and bent, fasten the carrying channel to the hanger wires (*Figure 12*). Loop the wires around the carrying channel at the point where the wires have been bent. Twist and fasten the wires using a saddle tie. It is important that the carrying channels are installed level. If it is necessary to extend the length of a carrying channel because of the room size, face the open U of the second channel toward the U of the one already installed. Overlap for a distance of 8 to 12 inches and tie together with wire (*Figure 13*).

Step 8 Move the dry line 4 feet in line with the next row of hangers. Continue to install the hanger inserts, wires, and carrying channels on 4-foot centers over the entire ceiling.

8.1.2 Installing Furring Members and Drywall (Gypsum Board)

After the carrying channel support members have been installed, proceed with the installation of the cross furring members. Different kinds of furring members can be used at right angles to the carrying channels at 16 inches on center (maximum). For example, a type called a hat furring channel is used to screw the drywall board to the furring channels with self-tapping screws. Install the furring channels as follows:

Step 1 Install furring channels perpendicular to the carrying channels. Place the first furring channel in line with the first perpendicular dry line. Attach the furring channel to the carrying channel using a furring channel clip.

Step 2 Continue to install the furring channels 16 inches on center. If it is necessary to extend the furring channels beyond their normal length, lap one end of the channel over the next piece a minimum of 8 inches and wrap the splice well with three-strand, 18-gauge tie wire (*Figure 14*). Avoid installing furring channels over places where fixtures will penetrate the ceiling.

Step 3 Screw the drywall to the furring channels (*Figure 15*). Be sure to use screws of sufficient length. The screw should be long enough to go through the board and extend through the furring channel about ⅝ inch.

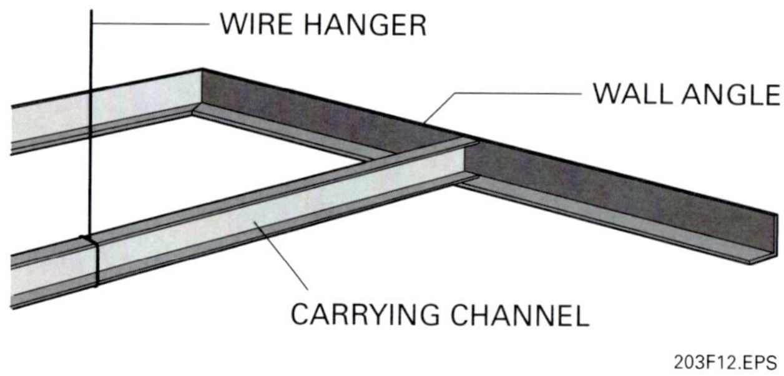

Figure 12 ♦ Carrying channel supported by wire hanger.

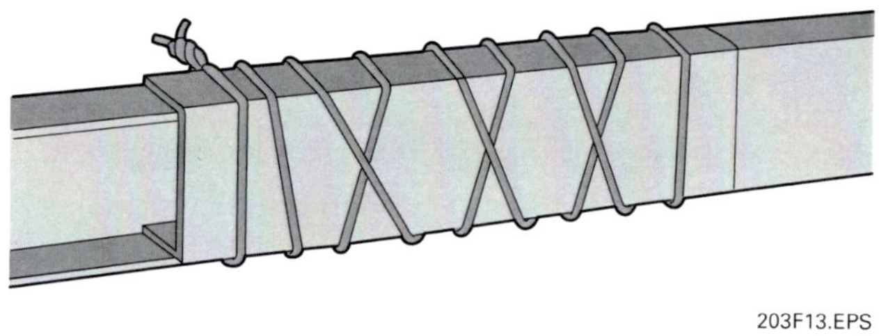

Figure 13 ♦ Splicing carrying channels.

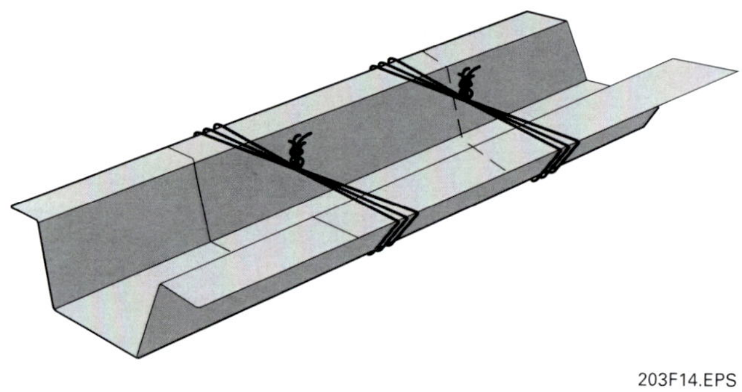

Figure 14 ♦ Splicing (lapping) furring channels.

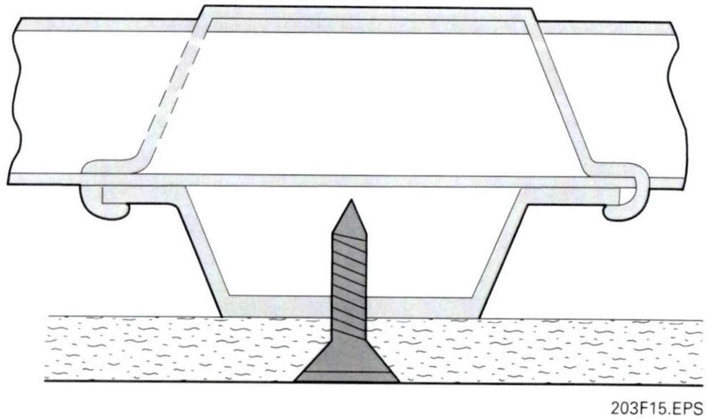

Figure 15 ♦ Drywall screwed into a furring channel.

8.1.3 Installing Acoustical Tile

Acoustical tile is used as a finish for drywall furring systems in cases where sound levels indicate a critical need for this tile. If acoustical tile is to be used as the finish over the drywall, proceed as follows:

Step 1 Lay out the room. Measure the length of one of the two shorter walls. Divide this figure in half. Mark this halfway point on the board next to this point on the wall. Next, measure the length of the facing wall at the opposite end of the room. Divide this wall length in half and mark this point on the board.

Step 2 Stretch a chalkline across the room and line it up with the two marks on the board. Snap the line.

Step 3 Divide the chalkline in half and set a second half-line perpendicular to the first one at this midpoint.

Step 4 Install the ceiling tile using the manufacturer's instructions starting at the junction of the two center lines.

8.1.4 Installing Furring Channels Directly to Structural Members

Sometimes furring channels are installed directly to structural members, such as open-web steel beams or wood joists, instead of to suspended carrying channels. In these instances, install the furring channels as follows:

Step 1 Mark the walls to show the height to be used to level the ceiling. Set the ceiling height control marks to indicate the position of the bottom edge of the furring channels. Check the level of the structural joists. If they are not level, use the low point of the joist as the common level for installing the furring channels.

Step 2 Measure up from the benchmarks to the height of the lowest point on the joists. Using the distance of the lowest point minus the height of the furring channel, mark it on all walls to indicate a uniform level for the lower edge of all furring channels. Be sure to measure up from the benchmarks to establish this common level on all walls.

Step 3 Establish the chalklines using the ceiling control marks.

Step 4 Install the wall moldings (if used). New chalklines may have to be snapped to guide the installation process.

Step 5 Hang the dry lines by securing them to facing walls at the chalkline level. Run the first line at this level right under the first joist. Run the second line at right angles to the first one at the same height. Position it to indicate where to install the first furring channel.

Step 6 Install the furring channels in the same manner as previously described, with the following exceptions:

- When attaching furring channels to steel joists, tie them together with the appropriate gauge tie wire. Wrap the wire around the two so that it bridges the joist and supports the furring channel on either side of the joists. Tie the wire ends together at the side of the union.
- Shimming may be needed between the joists and the furring channel. Be sure to check the level of the furring channel along its entire length. Shim where needed to correct any deviation from level.

8.2.0 Drywall Grid Systems

The drywall grid system (DGS) is a suspension ceiling system for drywall ceilings. In this system, a grid is suspended from the ceiling, and the drywall panels are installed in the grid. The drywall grid system makes hanging drywall ceilings

INSIDE TRACK

Special Furring Systems

Most manufacturers make furring system cross tees in standard 14-, 26-, and 50-inch lengths, as well as in other custom lengths. These sizes can reduce the time it takes to install an F-type lighting fixture from 30 minutes to less than one minute.

much easier and faster than the traditional method of nailing the drywall to the ceiling studs. This system is ideal for large ceilings, long flat ceilings, and floating ceilings. It can also be used in small areas, called short-span areas, such as hallways. In addition, DGS can be used to create dramatic, curved ceilings (*Figure 16*).

203F16.EPS

Figure 16 ◆ Curved drywall system.

An extruded aluminum drywall perimeter trim used in conjunction with the DGS provides a great finished edge for floating ceilings with straight or curved edges. Curved perimeter trim is factory pre-curved to the desired radius/radii. This trim will not only create a solution at the perimeter edge, but is a labor cost reducer and time saver that will ensure a better looking perimeter edge than drywall. The perimeter trim is secured to the drywall suspension with factory clips, screwed into the drywall and drywall suspension at the perimeter trim taping flange, and is taped and floated to the drywall for a seamless appearance. Clean and crisp perimeter edges are a guarantee if extruded aluminum trims are used. These trims are generally factory primed and field finished along with the drywall.

9.0.0 ◆ SPECIALTY SYSTEMS

Numerous specialty ceiling systems are available. These include metal pan systems, concealed grid systems, integrated ceiling systems, luminous ceiling systems, and others. Each offers specific design options, as described in the following sections.

9.1.0 Metal Pan Systems

The metal pan system is similar to the usual suspended acoustical ceiling system except that metal tiles or pans are used in place of the typical sound-absorbing panel (*Figure 17*).

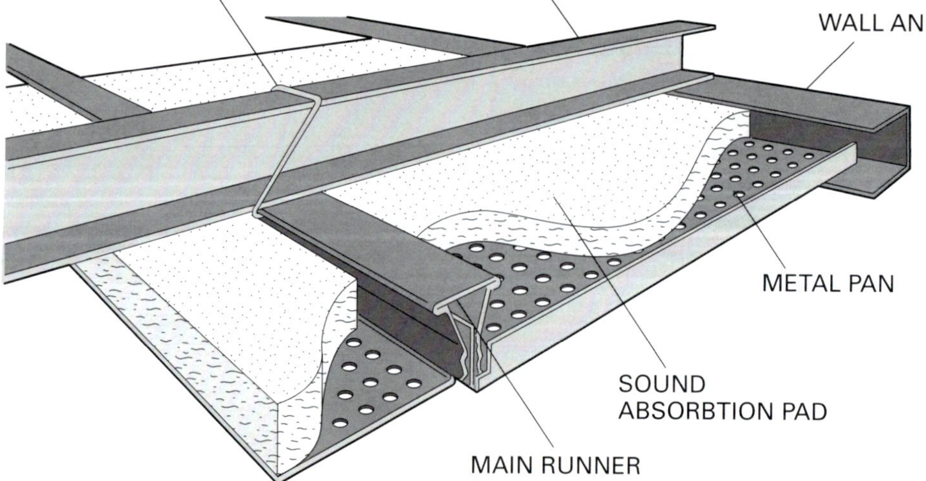

203F17.EPS

Figure 17 ◆ Metal pan ceiling components.

The pans are made of steel or aluminum. They are generally painted white, but other colors may be available. Pans are also offered in a variety of surface patterns. Tests have shown that metal pan ceiling systems are effective for sound absorption. They are durable and easily cleaned and disinfected. In addition, the finished ceiling has little or no tendency to have sagging joint lines or drooping corners. The metal pans are die-stamped and have crimped edges that snap into the spring-locking main runner to provide a flush ceiling.

The tools, room layout, and installation of hanger inserts, hangers, and wall angles for the metal pan system are basically the same as for the conventional exposed grid suspended ceiling.

Metal pan systems use 1½-inch U-shaped, cold-rolled steel furring channel members. They are normally installed 4 foot on center. However, it may be necessary to install them at lesser distances depending on the location of the light fixtures. The furring channels are installed by looping the hanger wires around them. Twist and secure them using saddle wire. It is important that the furring channels are properly leveled to ensure a level ceiling.

Once the furring channels have been installed, the main runners (tee bars) must be placed. These bars are manufactured, cold-rolled or zinc-coated steel or aluminum. They are fastened to the furring channels using furring clips. The tee bars run at right angles to the furring channels (*Figure 18*).

The tee bar has a spring-locking feature that grips the metal pan. This feature allows the pan to be removed for access to the area above the pan. If the room plan dimensions are longer than the length of the tee bar, use a tee bar splice to couple the tee bars together and extend them to the required length (*Figure 19*).

To install the tee bars, use the procedures that follow:

Step 1 Install the first tee bar at right angles to the 1½-inch furring channel. Be sure the bar is correctly aligned.

Step 2 When in position, place furring clips over the furring channel and insert the ends under the flange of the tee bar. Hang additional clips in the same manner along the length of the bar.

Step 3 Install the second tee bar parallel to the first one. The spacing should be from 1 to 4 feet, depending on the size of the metal pans.

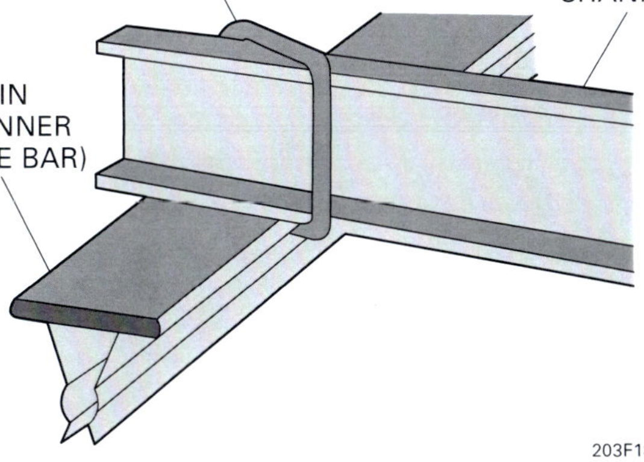

Figure 18 ◆ Main runner (tee bar) clipped to a furring channel.

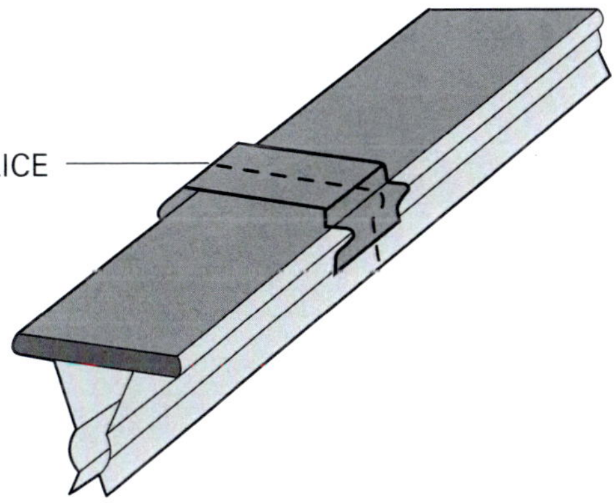

Figure 19 ◆ Tee bar splice.

After installing all the tee bars in one section of the room (depending on the working area of your scaffold), begin installing the metal pans. Take care when handling the pans. Use clean white gloves or wash your hands and then rub them with cornstarch to prevent any perspiration or grease marks from marring the surface of the pans. If care is not taken, fingerprints will be plainly visible when the units are installed. Always wash your hands before handling the pans and periodically throughout the job. Use the following procedures to remove the pans from their container:

Step 1 Place the pans finished-surface down on a raised platform (such as a table) that has been covered with a pad to protect the pan's surface.

Step 2 Insert the wire grid into the back of the metal pan. The wire grid is installed between the metal pan and backing pad to provide an air cushion between the two surfaces.

Step 3 Place the paper- or vinyl-wrapped mineral wool or fiberglass batt or pad over the grid.

> **NOTE**
> In some cases, the metal pans come with the wire grid and pad already assembled.

When installing the metal pans, start at the perimeter of the ceiling, next to the walls, since the units must fit into the channel wall angle. Use the following procedures:

Step 1 Obtain the width of the border units from the room layout or reflected ceiling plan. Measure the pan to this width.

Step 2 Using a band saw, cut the pan ⅛ inch short of the desired width along the edge to be inserted into the wall angle.

Step 3 Slide the pan into the wall angle. Do not force the pan all the way into the molding. Leave ⅛ inch for expansion.

Step 4 Insert two crimped edges of the metal pan into the spring-locking tee bars.

Step 5 When the pan is in position, insert the spacer clip into the channel molding cavity and over the cut edge of the pan. This will prevent the pan from buckling along this cut edge.

Step 6 Cut the wire grid and backing pad to fit the border unit. Place the unit into position.

Step 7 Install the pans at the corners of the room in the order shown in *Figure 20*.

Once the pans on the perimeter have been installed in each row, the full size pans can be put into place as follows:

Step 1 Grasping the pan at its edges, force the crimped edges into the tee bar slots. Use the palms of your hands to seat the pan.

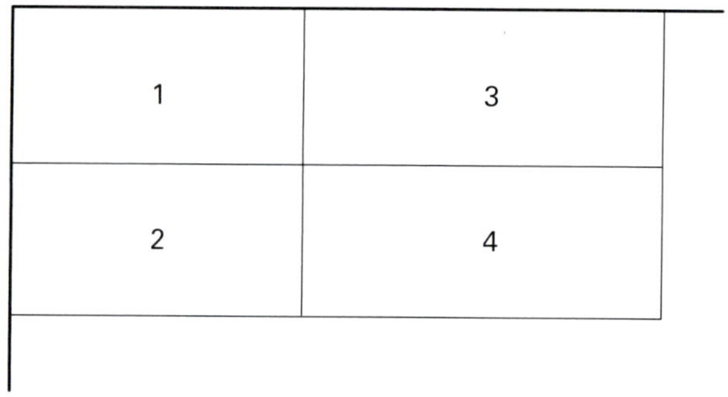

203F20.EPS

Figure 20 ◆ Panel layout.

Step 2 After installing several of the pans as noted above, slide them along the tee bars into their final positions. Use the side of your closed fist to bump the pans into level position if they do not seat readily.

Step 3 If metal pan hoods are required, slip them into position over the pans as they are installed. The purpose of the hood is to reduce the travel of sound through the ceiling into the room.

If a metal pan must be removed, a special tool is available (*Figure 21*). To remove a pan, start at one corner and insert the device into the joint between the pan to be removed and the adjacent pan. Push the device up until it engages the pan, and then gently pull down. Repeat this process at each corner of the pan. By following this removal procedure, there is no danger of bending the pan out of shape. *Figure 22* shows a finished metal pan ceiling.

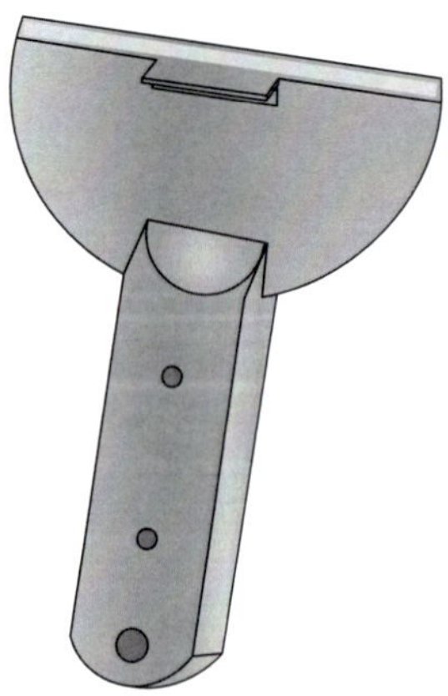

203F21.EPS

Figure 21 ◆ Pan removal tool.

203F22.EPS

Figure 22 ◆ Metal pan ceiling.

9.2.0 Direct-Hung Concealed Grid Systems

A concealed grid system is advantageous if a suspended ceiling system is to be used but the architect or designer desires the support runners to be hidden from view. This results in a ceiling that is not broken by the pattern of the runners (*Figure 23*).

The tiles used for this system are similar to standard tile, but are made with a kerf on all four edges. Kerfed and rabbeted 12" × 12" and 12" × 24" tiles are used with this system. Tiles of various colors and finishes are available. *Figure 24* shows the components of a typical concealed grid system.

Use these procedures to install a typical concealed grid system:

Step 1 The installation of the concealed grid system begins in the same way as the standard exposed grid system already discussed. Once chalklines have been established on the walls at the required height above the floor, the next step is to install the wall angle. This molding provides support for the grid and tile at the wall as it does for the other ceiling systems. The molding should be fastened with nails, screws, or masonry anchors. At corners, miter the inside corner and use a cap to cover the outside corner.

203F23.EPS

Figure 23 ◆ Concealed grid system.

INSIDE TRACK

Facts About Ceiling Panels

- Ceiling panels are often called pads.
- Certain lighting fixtures are manufactured in the same sizes as panels to drop directly into suspended ceilings.
- Some panels are treated to inhibit the growth of mold, mildew, fungi, and certain bacteria.
- Special panels are designed for use below sprinkler systems. These panels shrink and fall out of the grid as the room temperature rises, allowing the water to reach the fire.
- Most ceiling panels can be recycled.

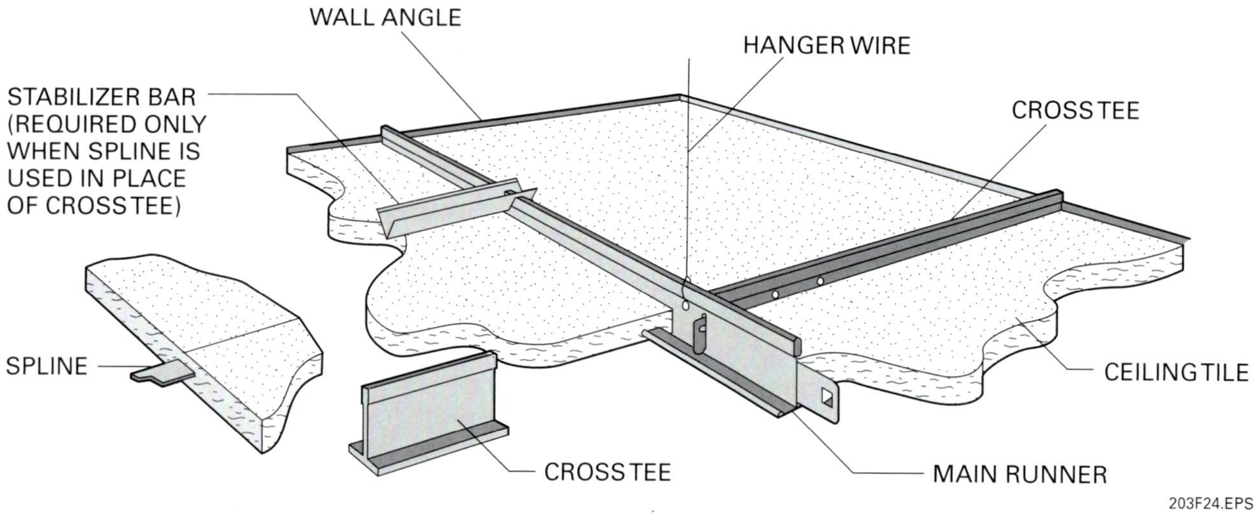

Figure 24 ♦ Direct-hung concealed grid system components.

Step 2 Lay out the grid and install the hanger inserts and hangers according to the reflected ceiling plan or lighting layout. Once this has been completed, install the concealed main runners as done for the exposed grid system. The main runners are the primary support members. The method of coupling them to attain a specific length will vary with the manufacturer of the system, but they can all be spliced to the desired length.

Step 3 Install the cross stabilizer bars and concealed cross tees at right angles to the main runners. They rest on the flange of the main tee runners.

Step 4 Place the ceiling tile into position. Cut in the first row of tile at a line perpendicular to the main runners, as previously established. Attach the tile to the wall using spring clips.

Flat anti-breather splines (*Figure 25*) are metal or fiber units that are inserted by the installer into the unfilled tile kerfs between the concealed main runners. Splines are used to prevent dust from seeping through the kerfs in adjacent tiles.

If access is needed to the area above the ceiling, special systems are available that can be incorporated into the ceiling (*Figure 26*).

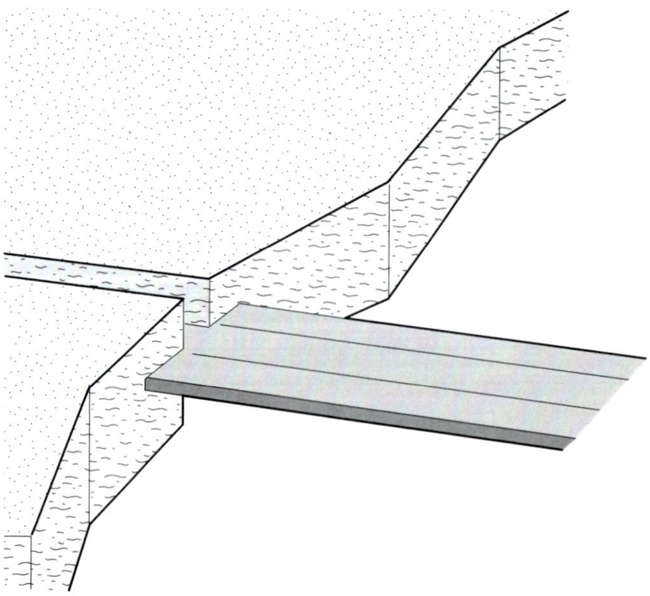

Figure 25 ♦ Flat anti-breather spline.

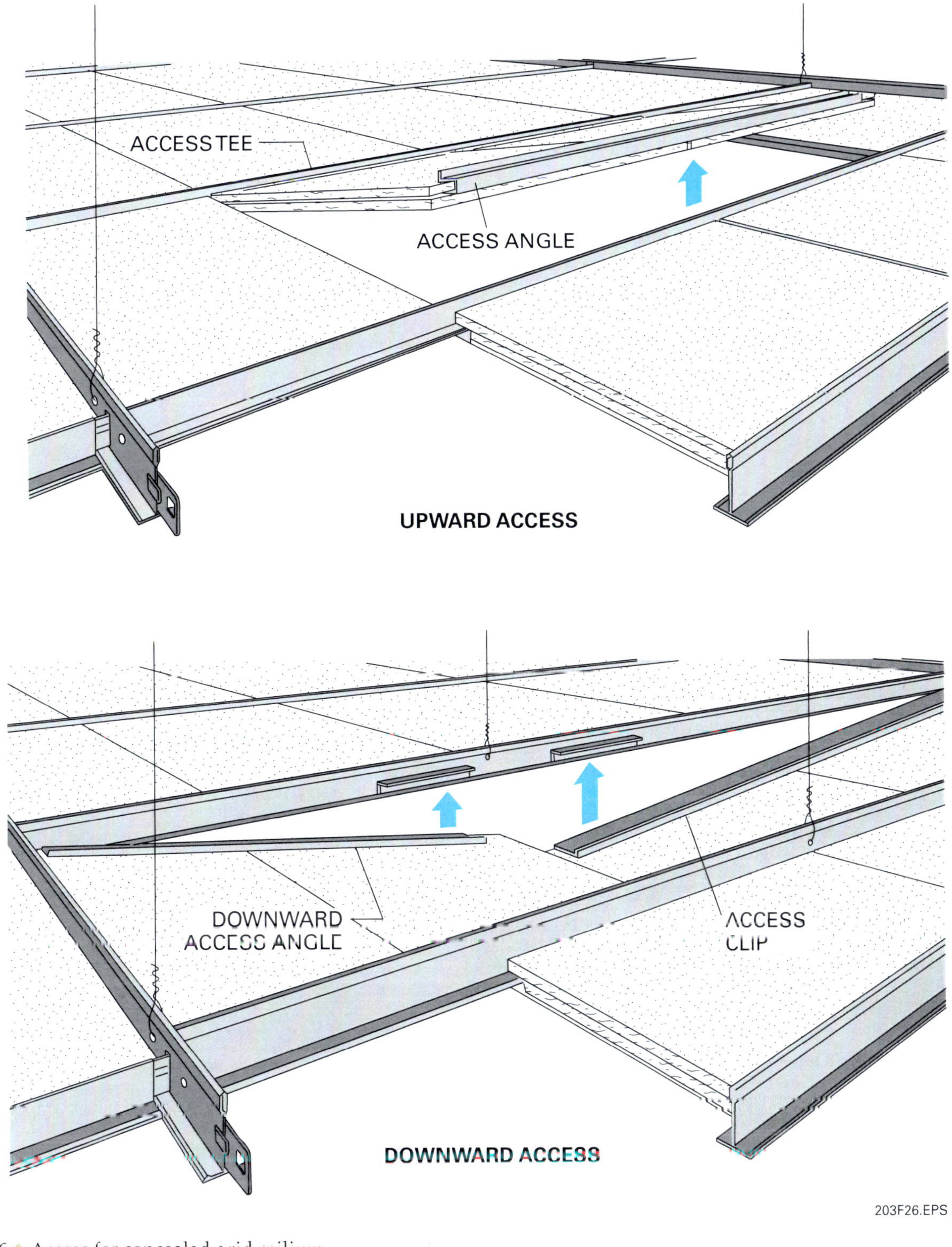

Figure 26 ♦ Access for concealed grid ceilings.

9.3.0 Integrated Ceiling Systems

The integrated ceiling system incorporates the lighting and/or air supply diffusers as part of the overall ceiling system, as shown in *Figures 27* and *28*.

Some manufacturers of ceiling materials offer specialty ceiling designs. Many integrated ceilings can be used to create distinctive architectural looks and artistic interior designs. Complete ceiling systems offer dozens of options, both functional and visually appealing.

The functional aspect allows for enhanced lighting and lighting effects. Mechanically, the air supply system can be incorporated through spaced supply air diffusers and return air grilles, all of which have been designed to go beyond function to enhance the artistic appearance of the ceiling.

Integrated ceiling systems are available in units called modules. Common sizes are 30" × 60" and 60" × 60". The dimensions refer to the spacing of the main runners and cross tees.

9.4.0 Luminous Ceiling Systems

Luminous ceiling systems (*Figure 29*) come in many styles, such as exposed grid systems with drop-in plastic light diffusers or in an aluminum or wood framework with translucent acrylic light diffusers.

Fluorescent fixtures are generally installed above the translucent diffusers. Standard modules are available of 2' × 2' up to sizes of 5' × 5'. It is also possible to purchase custom sizes for special conditions. The two types of luminous ceilings are standard and non-standard. Standard systems are available in a series of standard sizes and patterns. Non-standard systems deviate from the normal spacing of main supports and may include unusual panel sizes, shapes, and configurations.

All surfaces in the luminous space, including pipes, ductwork, ceilings, and walls, are painted with a 75 to 90 percent reflectance matte white finish. Any surfaces above the luminous ceiling system that might flake, such as fire proofing and insulation, should receive an approved hard sur-

203F27.EPS

Figure 27 ♦ Integrated grid system.

203F29.EPS

Figure 29 ♦ Luminous ceiling system.

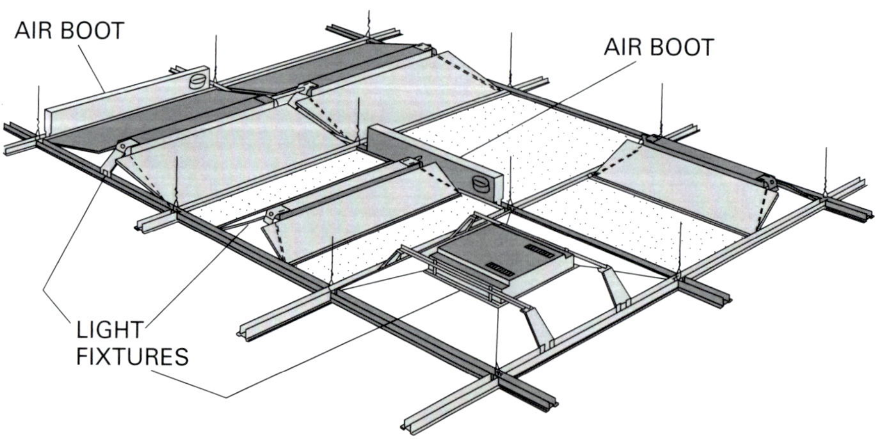

203F28.EPS

Figure 28 ♦ Integrated ceiling schematic.

face coating prior to painting. This will prevent flaking onto the ceiling below.

9.4.1 Installation of Standard Systems

A standard luminous system is installed in the same way as an exposed grid suspended system, except for the border cuts. Luminous ceilings are placed into the grid members in full modules. Any remaining modules are filled in with acoustical material that has been cut to size.

With a 2' × 2' or 2' × 4' standard exposed grid system, use luminous panels, which provide the light diffusing element in the system. These panels are laid in between the runners. Many panel sizes and shapes are available.

9.4.2 Installation of Non-Standard Systems

A non-standard luminous ceiling is installed in the same way as a standard system with regard to room layout, hanger insert, installation of hanger wire, main supports, secondary supports, and attachment of some type of wall angle.

Since non-standard ceilings can differ so much in terms of size, complexity of the system, and the exactness of the installation, shop drawings are required. More information on non-standard ceilings can be obtained from ceiling manufacturers.

9.5.0 Other Specialty Ceiling Systems

There are many specialty ceiling systems that differ from those covered in this module. Some of these include special metallic systems, wood systems, planar systems, and reflective (mirrored) systems (*Figures 30* through *33*).

203F30.EPS

Figure 30 ♦ Metallic system.

GOING GREEN

Reflective Ceilings

Reflective ceiling panels can reduce lighting costs because they allow the use of lower-wattage light bulbs.

203F31.EPS

Figure 31 ♦ Pan system.

203F32.EPS

Figure 32 ♦ Planar system.

203F33.EPS

Figure 33 ♦ Reflective ceiling.

Transparent panels produce a subtle, refracted light that extends over all areas of a room's interior (*Figure 34*).

203F34.EPS

Figure 34 ◆ Transparent panel system.

10.0.0 ◆ SEISMIC CONSIDERATIONS FOR CEILINGS

Although important in their function, modern ceilings are generally not considered part of a building's structural system. Ceilings today are used to make rooms and buildings more livable. The most desirable features of a ceiling are those that improve the acoustic properties of an area and those that make it easier to heat, cool, and light. Although ceilings are not structurally significant, the International Building Code (IBC) recognizes that ceiling failures due to seismic activity can make an area unusable.

Although seismic activity is of special importance in earthquake-prone areas, such as the Pacific Northwest region of the United States, it is also an issue in other areas. The IBC uses three factors that place more than half of the United States in areas at risk for seismic activity. The factors the IBC uses for determining a building's seismic risk are as follows:

- *Ground movement* – All geographic areas have the potential for ground movement, even if it is unlikely that the movement will be felt. The primary factor is the amount of potential ground movement. In the United States, the state with the potential for the most ground movement is California, while the ones with the least potential are Florida and Texas.
- *Soil classification* – Soil classifications are based on the soil type at the building site to a point 100 feet below the surface. For seismic purposes, there are six soil classifications, labeled A through F. Class A is solid rock, which is considered the most stable. Class F soil has no positive construction traits and is unusable for building. It is unlikely that a site would be classified at the extremes of A or F. Most sites will fall somewhere in between.
- *Building use* – Building use falls into one of three groups. Group 1 buildings have low occupancy, such as small residential buildings. Group 2 buildings have high occupancy—shopping malls and large office buildings are examples. Group 3 buildings are used for essential services, such as law enforcement and medical services. Group 3 is also used to describe those buildings that are normally classified as Group 1 or 2 but are three stories or more in height.

Based on the factors of potential ground movement, soil type, and building use, a Seismic Design Category has been established. The categories range from A to F, with category A assigned to buildings with the least seismic risk. Buildings in categories A and B use the standard suspended ceiling requirements described in this module. Their ceilings are anchored to walls and suspended from the upper deck using standard hardware. Those buildings in category C should have unrestrained ceilings that are free-floating to allow for movement of up to 12 inches and 45 degrees without damage. Buildings in categories D, E, and F should have restrained ceilings that are reinforced with stabilizing bars, heavier gauge wire, and multiple strands of wire.

The ceilings described in categories C through F are more expensive than those in categories A and B. Unrestrained ceilings (used in category C buildings) are not anchored to walls, so they are extremely difficult to install without a fixed reference. Unrestrained ceilings cost about 50 percent more than standard suspended ceilings—mostly in installation costs. Unrestrained ceilings have the following features:

- The ceiling system grid is not attached to the wall in any way.
- There are no perimeter wires.
- The wall molding must be at least ⅞ inch in width.

- The clearance between the ceiling and walls is at least ⅜ inch at all points.
- Grids joints must overlap at least ⅜ inch.
- The main beams and cross tees ends, which are free-floating, must be tied together to prevent spreading.

Restrained ceilings (used in categories D, E, and F) are anchored to at least two adjacent walls and require the use of heavy gauge hardware, as well as a grid system. These ceilings usually cost twice as much as a standard suspension ceiling, with materials accounting for most of the increased cost. Restrained ceilings (*Figure 35*) have the following characteristics:

- A heavy-duty ceiling grid system is required.
- The grid must be attached to two adjacent walls, and the clearance between the ceiling and the opposite walls must be ¾ inch.
- The wall molding must be at least 2 inches in width.
- The ceiling perimeter must have support wires that are spaced not more than 8 feet from the wall.
- Ceilings with an area of less than 1,000 square feet must use heavy-duty hardware for equipment mounted through the ceiling, such as light fixtures and sprinklers.
- Ceilings with an area of more than 1,000 square feet must have additional bracing.
- Ceilings with an area of more than 2,500 square feet must have seismic separation joints.
- Cable trays and electrical conduits must have their own supports and braces.

11.0.0 ◆ LAYING OUT AND ESTIMATING MATERIALS FOR A SUSPENDED CEILING

The estimate of materials for a suspended ceiling should be based on the ceiling plan provided with the architect's drawing set or on a scaled sketch of the ceiling layout. These drawings should show the direction and location of the main runners, cross tees, light panels, and border panels. In a typical suspended ceiling, the main runners are spaced 4 feet apart and are usually run parallel with the long dimension of the room.

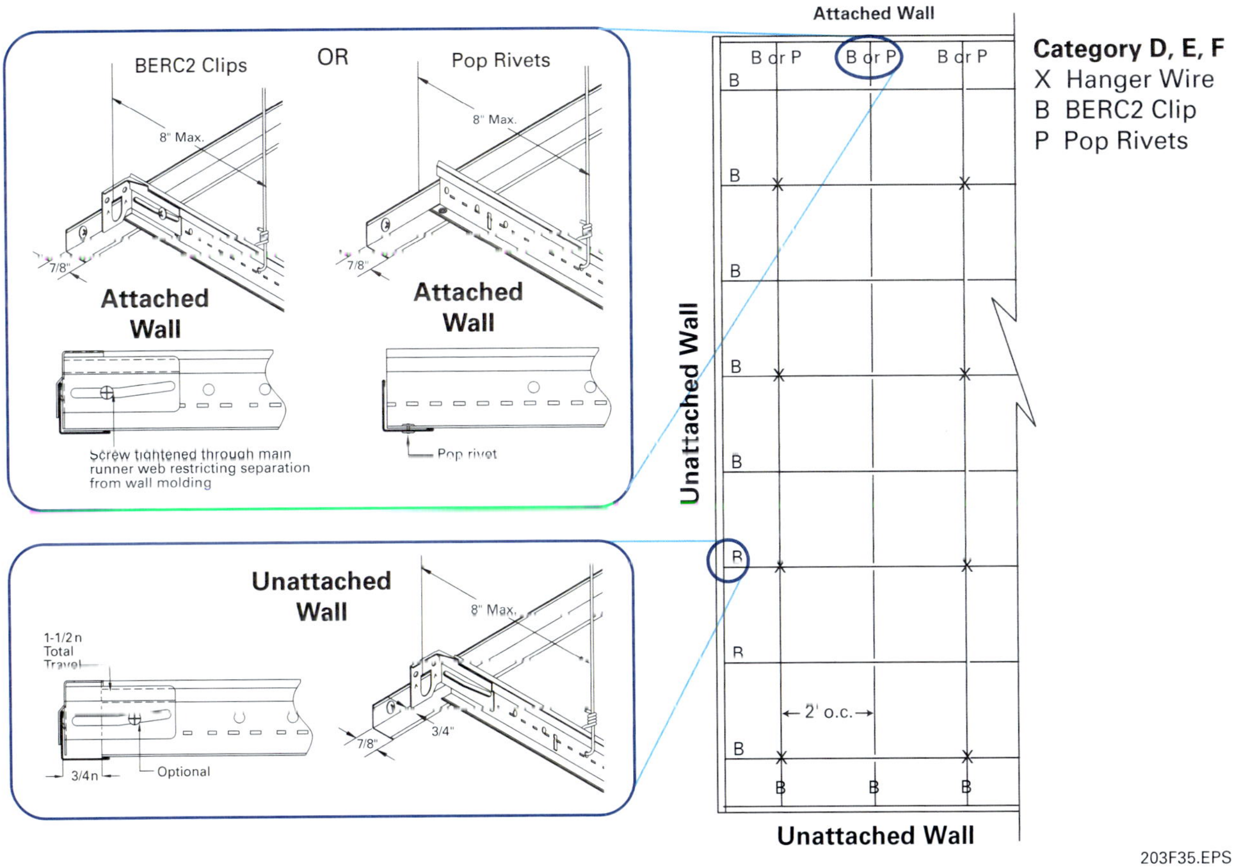

Figure 35 ◆ Seismic category D, E, and F ceiling requirements.

For a standard 2' × 4' pattern, 4-foot cross tees are spaced 2 feet apart between the main runners. If a 2' × 2' pattern is used, 2-foot cross tees are installed between the midpoints of the 4-foot cross tees. Main runners and cross tees should be located in such a way that the border panels on both sides of the room are equal and are as large as possible.

If no ceiling plan or sketch is available, you will need to make one to determine the required quantity of materials. Sketch a ceiling plan to scale on graph paper. Use a convenient scale, for example, one square equals one square foot. Measure along the ceiling level, including irregular areas such as bays, alcoves, and beams. Note each dimension on the drawing, then proceed as follows:

Step 1 Determine and sketch the locations of the main runners. Perform the following related procedures:

- Convert the dimension for the room width to inches.
- Divide the room width in inches by 48" and add 48" to any remainder.
- Divide the sum by 2 to find the distance from the wall to install the first main runner. Note that this distance is also the length of the border panels or tiles. For example, assume the room has a width of 12'-8" and a length of 20'-8", as shown in *Figure 36*. Changing the room width to inches equals 152". Dividing 152" by 48" yields 3, with a remainder of 8". Adding 48" to 8" equals 56". Dividing 56" by 2 equals 28", the distance from the wall to the first runner. If there is no remainder, the distance from the wall to the first runner is 48".
- Draw a main runner the calculated distance from, and parallel to, the long dimension of the ceiling. Draw the remaining main runners parallel to the first runner at 4 feet on center. The distance between the last main runner and the wall should be the same as the distance between the first main runner and the wall.

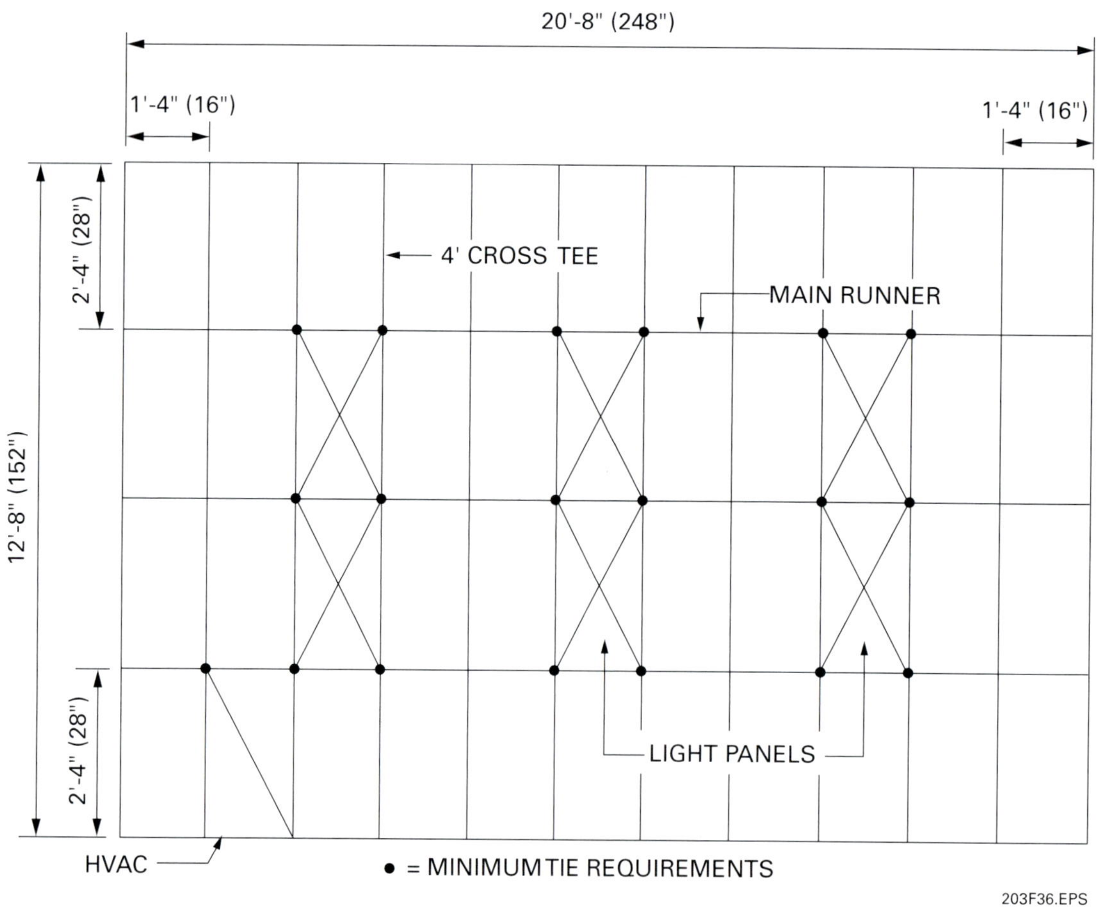

Figure 36 ◆ Completed sketch of a suspended ceiling layout.

Step 2 Determine and sketch the locations of the 4-foot cross tees.

- Convert the long dimension of the room to inches.
- Divide the room length in inches by 24" and add 24" to any remainder.
- Divide the sum by 2 to find the distance in inches from the wall to install the first row of cross tees. Note that this dimension is also the width of the border panels or tiles. For our sample room, the length is 20'-8". Changing this dimension to inches equals 248". Dividing 248" by 24" yields 10, with a remainder of 8". Adding 24" to 8" equals 32". Dividing 32" by 2 equals 16", the distance from the wall to the first row of cross tees.
- Draw the first row of cross tees the calculated distance from, and parallel to, the short wall. Draw the remaining rows of cross tees parallel to each other at 2 feet on center. The distance from the last row of cross tees to the wall should be the same as the distance from the first row of cross tees to the opposite wall.

NOTE

Additional 2-foot cross tees are required if 2' × 2' panels are used. You will also need one hanger device and wire for about every 4 feet of main runner.

Step 3 Determine the quantity of ceiling materials needed. From the ceiling plan or sketch, determine the number of pieces required for the wall angle, main runners, cross tees, and ceiling panels. Remember that decimals and fractions need to be rounded up when calculating material needs. Normally, main runners come in 12-foot lengths, cross tees in 4-foot lengths, and wall angle in 10- or 12-foot lengths depending on the type.

Step 4 Using the ceiling plan or sketch, find the number of main runner sections needed. For our sample room (*Figure 36*), 6 main runner sections are required. This is because main runners are made 12 feet in length and no more than two pieces can be cut from any one 12-foot runner. Three main runners, each 20'-8" (20.667') in length, total 62.001'. Therefore, 62.001' ÷ 12' = 5.167 lengths or 6 lengths when rounded up.

Step 5 Find the number of 4-foot cross tees. For our sample room, 40 cross tees (10 per row × 4 rows) are required. Note that the border cross tees must be cut from full length cross tees.

Step 6 If using 2' × 2' panels or tiles, find the number of 2-foot cross tees. A 2' × 2' grid is made by installing 2-foot cross tees between the midpoints of the 4-foot cross tees. The number of 2-foot cross tees required for the sample room is 44 (11 per row × 4 rows).

Step 7 Find the number of sections of wall angle needed. Divide the perimeter of the room (perimeter = [2 × length] + [2 × width]) by 10 feet. For our sample room, the perimeter is 66'-8" (66.667'). Therefore, 7 sections are needed (66.667' ÷ 10' = 6.667 or 7 when rounded up).

Step 8 Find the number of ceiling panels or tiles. One method is to count the total number of ceiling panels shown on the ceiling plan or sketch. Note that each border panel requires a full size ceiling panel. Also, subtract one panel for each lighting fixture installed in the ceiling. Assuming the use of 2' × 4' ceiling panels and 6 light fixtures, our sample ceiling would require 38 panels.

Step 9 Find the approximate number of hanger wires and hangers needed. Assume one hanger device and wire for about every 4 feet of main runner. For our example ceiling, approximately 16 hangers are required (62.001' ÷ 4' = 15.5 or 16 when rounded up). Multiply the number of hangers needed by the required length of each hanger wire to find the total linear feet of hanger wire needed.

NOTE

Another way to determine the total square footage of the ceiling is by multiplying the length times the width (area = length × width). Then, divide the total ceiling area by the coverage in square feet printed on the carton of the panels intended for use. If using panels that cover 64 sq ft per carton, the example room ceiling would require 4.09 cartons of ceiling panels (12.667' × 20.667' = 261.789' ÷ 64 = 4.09 or four cartons plus one extra panel).

11.1.0 Alternate Method for Laying Out a Suspended Ceiling Grid System

The following steps describe another common method for laying out the grid system for a suspended ceiling:

Step 1 Locate the room center line parallel to the long dimension of the room and draw it on the ceiling sketch.

Step 2 Beginning at the center line and going toward each side wall, mark off 4-foot intervals on the sketch. If more than a 2-foot space remains between the last mark and the side wall, locate the main runners at these marks. If less than a 2-foot space remains between the last mark and the side wall, locate the main runners at 4-foot intervals beginning at 2 feet on either side of the center line. This procedure provides for symmetrical border panels of the largest possible size. Remember to consider the locations of light fixtures and air diffusers in the room when planning the ceiling layout.

Step 3 Locate the 4-foot cross tees by drawing lines 2 feet on center at right angles to the main runners. To obtain border panels of equal size, begin at the center of the room using the same procedure described in Step 2.

Step 4 If using a 2' × 2' grid pattern, locate the 2-foot cross tees by bisecting each 2' × 4' module.

Step 5 In the same manner as was described previously, estimate the materials for the grid system using the information shown on the ceiling sketch.

12.0.0 ◆ CEILING CLEANING

Immediately after installation and after extended periods of use, some suspended ceilings may require cleaning to maintain their acoustical performance characteristics and attractiveness. Dust and loose dirt can easily be removed with a brush or vacuum cleaner. Vacuum cleaner attachments, such as those designed for cleaning upholstery or walls, do the best job. Make sure to clean in one direction only to prevent rubbing dust or dirt into the surface of the ceiling.

After loose dirt has been removed, pencil marks, smudges, or clinging dirt can often be removed with an ordinary art gum eraser. Most mineral fiber ceilings can be cleaned with a moist cloth or sponge. The sponge should contain as little water as possible. After washing, the soapy film should be wiped off with a cloth or sponge slightly dampened in clean water. Vinyl-faced fiberglass ceilings and mylar-faced ceilings can be cleaned with mild detergents or germicidal cleaners.

> **INSIDE TRACK**
>
> **Cleaning Ceiling Panels**
>
> Improper cleaning can damage ceiling panels. The materials and methods used to clean them vary widely, depending on the finish and texture of the panels. One manufacturer prescribes eight different cleaning methods for their line of ceiling panels. It is very important to check the manufacturer's instructions before proceeding. When cleaning grids, the ceiling panels should be removed to prevent cleaning solution or dirt from getting on the panels.

13.0.0 ◆ INSTALLATION BEST PRACTICES

Following these guidelines will help you to achieve the desired level of professionalism when installing ceilings:

- Make sure that all workmanship meets the highest standards in accordance with the Ceiling and Interior Systems Contractors' Association policy of upgrading quality.
- Arrange ceiling panels so that units less than one-half width do not occur unless otherwise directed by the reflected ceiling plans or job conditions.
- Make sure that all panels, tile joints, and exposed suspension systems are straight and in alignment.
- All acoustical ceiling systems must be level to $\frac{1}{8}$ inch in 10 feet.
- Tile must be neatly scribed against butting surfaces and to all penetrations or protrusions where moldings are not required.
- Install tiles surrounding recessed lights and similar openings with a positive method to prevent movement or displacement of the tiles.
- Install tiles in a uniform manner with neat hairline-fitted joints between adjoining tiles.
- Make sure that wall moldings are firmly secured, the corners neatly mitered, or corner caps used, if preferred.
- The completed ceiling must be clean and in undamaged condition.

Review Questions

1. Sound waves travel _____ from their source.
   a. on a line-of-sight
   b. in all directions
   c. vertically
   d. at right angles

2. The intensity of a sound wave refers to _____.
   a. the number of wave cycles it completes in a second
   b. its loudness
   c. its frequency
   d. both its loudness and softness

3. A sound level of 100 dBA is considered _____.
   a. very loud
   b. moderately loud
   c. loud
   d. quiet

4. The term used by manufacturers to compare the noise absorbency of their materials is the _____.
   a. ceiling attenuation class (CAC)
   b. articulation class (AC)
   c. noise reduction coefficient (NRC)
   d. sound transmission classification (STC)

5. On most large construction jobs, there is often _____ ceiling plan supplied in the plan set.
   a. a projected
   b. a reflected
   c. a reversed
   d. no

6. In an exposed grid ceiling system, hanger wire is used to support the _____.
   a. 4-foot cross tees
   b. 2-foot cross tees
   c. wall angle
   d. main runners

7. An exposed grid ceiling system uses the _____ as the primary support members of the grid system.
   a. main runners
   b. 4-foot cross tees
   c. wall angle
   d. 2-foot cross tees

8. Acoustical panels and tiles used in suspended ceilings stop the transmission of unwanted sounds by _____ the sound waves.
   a. reflecting
   b. reverberating
   c. absorbing
   d. refracting

9. When laying out ceiling clips for hanger inserts, two dry lines are used to determine the _____.
   a. exact center of the room
   b. width of the border panels
   c. center line of the first ceiling panel
   d. location of the first ceiling clip

10. When installing furring channel members for a suspended drywall ceiling system directly to steel beams or wood joists above the ceiling, _____ can be used against the beams or joists as needed to correct any deviations from level.
    a. shims
    b. spacers
    c. main runners
    d. leveling rods

11. When installing the metal pans in a metal pan ceiling system, begin by installing them _____.
    a. in the middle of the ceiling
    b. at the perimeter of the ceiling next to the walls
    c. at the corners of the room
    d. at the most convenient location

Review Questions

12. Splines installed into the unfilled tile kerfs in a direct-hung concealed grid ceiling system _____.
    a. prevent buckling of the ceiling
    b. are used in place of stabilizer bars
    c. prevent shifting of the tiles
    d. prevent dust from seeping through the kerfs in adjacent tiles

13. The size of panel (module) that is commonly used in an integrated ceiling system is _____.
    a. 12" × 12"
    b. 12" × 24"
    c. 24" × 24"
    d. 30" × 60"

14. All ceilings, walls, pipe, and ductwork in the space above a luminous ceiling will normally be painted with a _____ paint.
    a. light refracting
    b. 75 to 90 percent reflective matte finish
    c. 90 to 100 percent reflective matte finish
    d. luminous matte finish

15. The Seismic Design Category for a building is determined by _____.
    a. soil type, building design, and potential ground movement
    b. building use, soil type, and building design
    c. potential ground movement, building design, and building use
    d. building use, soil type, and potential ground movement

16. With a suspended ceiling that is 10' × 16', at what distance from the wall will the first main runner be installed if 2' × 4' ceiling panels are to be used with the main runners parallel with the ceiling's long dimension and the 4-foot length of the panels parallel with the ceiling's short dimension?
    a. 18 inches
    b. 24 inches
    c. 36 inches
    d. 48 inches

Questions 17 through 20 refer to the ceiling described in Question 16.

17. How many 12-foot main runners are required for the ceiling?
    a. 2
    b. 3
    c. 4
    d. 5

18. How many 4-foot cross tees are required for the ceiling?
    a. 16
    b. 20
    c. 21
    d. 22

19. Assuming two light fixtures are used for the ceiling, how many 2' × 4' ceiling panels are required?
    a. 18
    b. 20
    c. 22
    d. 24

20. What are the dimensions (width and length) of the border ceiling panels to be installed along the long dimension of the ceiling?
    a. 18" × 24"
    b. 24" × 30"
    c. 24" × 36"
    d. 36" × 48"

Summary

This module provided information about the major types of acoustical suspended ceiling systems and the basic procedures for installing them. Although other trades may be responsible for ceiling installations on some jobs, it will likely be the drywall technician's responsibility to do the work. For this reason, it is important that you understand the fundamentals of acoustical suspended ceiling systems and their installation. For more detailed information about specific products and their use, consult the manufacturer's instructions.

Notes

Trade Terms Introduced in This Module

Acoustical materials: Types of ceiling tile, plaster, and other materials that have high absorption of sound waves.

Acoustics: A science involving the production, transmission, reception, and effects of sound. In a room or other location, it refers to those characteristics that control the reflection of sound waves and the sound reception in the area.

A-weighted decibel (dBA): A single number measurement based on the decibel but weighted to approximate the response of the human ear to the intensity of sound sources.

Ceiling panels: Any lay-in acoustical board that is designed for use with an exposed grid mounting system. Ceiling panels normally do not have finished edges or precise dimensional tolerances because the exposed grid mounting system provides the trimout.

Ceiling tiles: Acoustical ceiling boards that are nailed to, cemented to, or suspended from a concealed grid mounting system. The edges are often kerfed and cut back.

Diffuser: An attachment for duct openings in air distribution systems that disperses the air in wide flow patterns. In lighting systems, it is an attachment used to redirect or scatter the light from a light source.

Fissured: A ceiling panel or ceiling tile surface design that has the appearance of splits or cracks.

Frequency: Cycles per unit of time, usually expressed in hertz (Hz).

Hertz (Hz): A unit of frequency equal to one cycle per second.

Striated: A ceiling panel or ceiling tile surface design that has the appearance of fine parallel grooves.

Additional Resources and References

Additional Resources

This module is intended to be a thorough resource for task training. The following reference works are suggested for further study. These are optional materials for continued education rather than for task training.

CISCA Ceiling Systems Handbook. St. Charles, IL: The Ceiling and Interior Systems Contractors' Association, 1999.

The Gypsum Construction Handbook. Chicago, IL: USG Corporation.

Figure Credits

Photos courtesy of constructionphotographs.com, 203F01, 203SA01, 203SA02

Ritterbush-Ellig-Hulsing P.C., 203F05

CST/berger, 203F06

Gordon Inc./www.gordonceilings.com, 203F16

Armstrong World Industries, 203F22, 203F23, 203F30

BPB Capaul Brand Ceilings, 203F27

Ceilings Plus, 203F29, 203F33, 203F34

Chicago Metallic Corporation, 203F31, 203F32

NCCER CURRICULA — USER UPDATE

NCCER makes every effort to keep its textbooks up-to-date and free of technical errors. We appreciate your help in this process. If you find an error, a typographical mistake, or an inaccuracy in NCCER's curricula, please fill out this form (or a photocopy), or complete the online form at **www.nccer.org/olf**. Be sure to include the exact module ID number, page number, a detailed description, and your recommended correction. Your input will be brought to the attention of the Authoring Team. Thank you for your assistance.

Instructors – If you have an idea for improving this textbook, or have found that additional materials were necessary to teach this module effectively, please let us know so that we may present your suggestions to the Authoring Team.

NCCER Product Development and Revision

13614 Progress Blvd., Alachua, FL 32615

Email: curriculum@nccer.org
Online: www.nccer.org/olf

❑ Trainee Guide ❑ Lesson Plans ❑ Exam ❑ PowerPoints Other ______________________

Craft / Level: Copyright Date:

Module ID Number / Title:

Section Number(s):

Description:

Recommended Correction:

Your Name:

Address:

Email: Phone:

45204-09

Interior Specialties

45204-09
Interior Specialties

Topics to be presented in this module include:

Overview

The number and variety of interior finish products is almost endless. In commercial buildings, finish products need to be attractive, easy to install and maintain, and should support the desired acoustical qualities of an area. Most finish products used in commercial buildings incorporate some type of design or material that improves an area's acoustical properties. This module provides information about some of the most frequently used wall and ceiling systems, including fabric-covered panels, wall upholstery, and wood panels and slats. General installation procedures are also covered. Unique finish products, such as ceiling domes and architectural features, are briefly discussed.

Objectives

When you have completed this module, you will be able to do the following:

1. Describe the composition and use of common types of specialty interior products.
2. Identify the different installation and attachment methods used with specialty interior products.

Trade Term

Veneer

Prerequisites

Before you begin this module, it is recommended that you successfully complete *Core Curriculum; Drywall Level One;* and *Drywall Level Two,* Modules 45201-09 through 45203-09.

This course map shows all of the modules in the second level of the *Drywall* curriculum. The suggested training order begins at the bottom and proceeds up. Skill levels increase as you advance on the course map. The local Training Program Sponsor may adjust the training order.

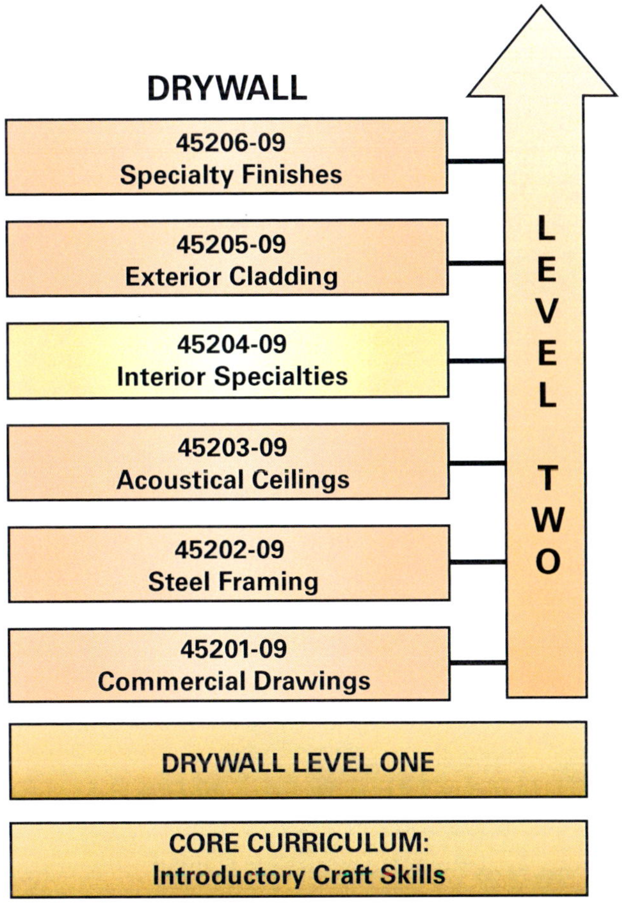

204CMAP.EPS

1.0.0 ◆ INTRODUCTION

Interior finish work makes commercial areas attractive and more usable. The number of interior finish products available is staggering. Although looks are important, the most important characteristic of any large space is its acoustics. In the past, a great deal of time and effort went into interior finish work, and much of it was done by hand. Today, an interior finish system must make an area attractive and improve its acoustics, but it also should be easy to install.

Manufacturers spend a great deal of time and money designing systems that are easy to install and maintain while looking attractive and unique. However, the most important feature of an interior finish system is its acoustic properties. Systems can often be used as both ceilings and wall coverings. Keep the following points in mind when installing any finish system:

- Neatness counts. Before working with any system, sweep or vacuum the installation area.
- Protect the finish system from dirt, scratches, and other damage.
- Handle materials with care to avoid damaging their acoustic properties and finish.
- Store all finish materials in a clean, safe location.
- Wash your hands before handling any easily soiled materials.

2.0.0 ◆ VINYL- AND FABRIC-COVERED FIBERGLASS PANELS

Fiberglass panels (*Figure 1*) are mounted on walls and ceilings to change the acoustical properties of a room. These panels are often used in large, open areas that would normally reflect and echo sounds. Areas such as auditoriums and churches, or places that are often crowded with people in conversation, such as reception rooms, can often benefit from these panels. Panels are available in many sizes and shapes, and are covered with fabric or vinyl to complement the decorating scheme of an area.

Fabric-covered wall panels can be easily damaged, so be careful when handling them and follow the basic guidelines provided in this section.

Store panels in an inside area on the job site where they will be protected from moisture and humidity. Always store panels flat and covered to protect the fabric from dust and dirt.

Carry panels upright and by their edges to avoid bending them during installation. Use a two-person carry or a sheetrock cart, if necessary.

DID YOU KNOW?

29 Years Later

The first president to live in the White House was John Adams. In November of 1800, he and his wife, Abigail, moved into the unfinished building. Mrs. Adams used what we now call the East Room to hang laundry to dry! It wasn't until 1829 that the East Room was finished and furnished.

204F01.EPS

Figure 1 ◆ Fabric-covered fiberglass wall panels.

It is important to protect the fabric or vinyl during installation. Wash your hands and wear clean white gloves to keep the panels unsoiled. Do not install damaged panels.

Be sure to clean the walls and ceiling of dust, dirt, or grease before installing panels with adhesive. Scrape any loose or peeling paint from the wall's surface.

Plan the installation of the panels before you begin. Snap vertical and horizontal chalklines on the walls as installation guides.

When panels must be cut to fit an area, pull the fabric away from the panels, and then cut the panels to size with a sharp blade or carpet knife. Trim the fabric as necessary, and reattach it to the panels by spraying adhesive on the back of the panels. You may need to trim a small wedge of fabric from the corner to get the fabric smoothly reattached to the panel.

> **NOTE**
>
> Sometimes a strip of plastic laminate is fastened with spray adhesive on the cut edge of the panel to sharpen the edge and make the fabric fold look more attractive.

When you must cut a hole in a panel to accommodate a fixture, be very careful not to cut the fabric when you are cutting the panel. To protect the fabric, pull it from one side of the panel, and then carefully place a piece of plastic, cardboard, or wood under the area to be cut. After cutting the desired shape from the panel, cut the fabric in an X-pattern, and then smoothly fold the triangular-shaped pieces over the back of the panel. For circular cut-outs, cut the fabric in two X-shaped patterns to create eight equal segments, as shown in *Figure 2*. You may want to apply a thin line of adhesive to the edge of fabrics that tend to unravel. When you are finished, attach the fabric to the back of the panel with spray adhesive.

There often will be detailed installation drawings available to show installation of the panels.

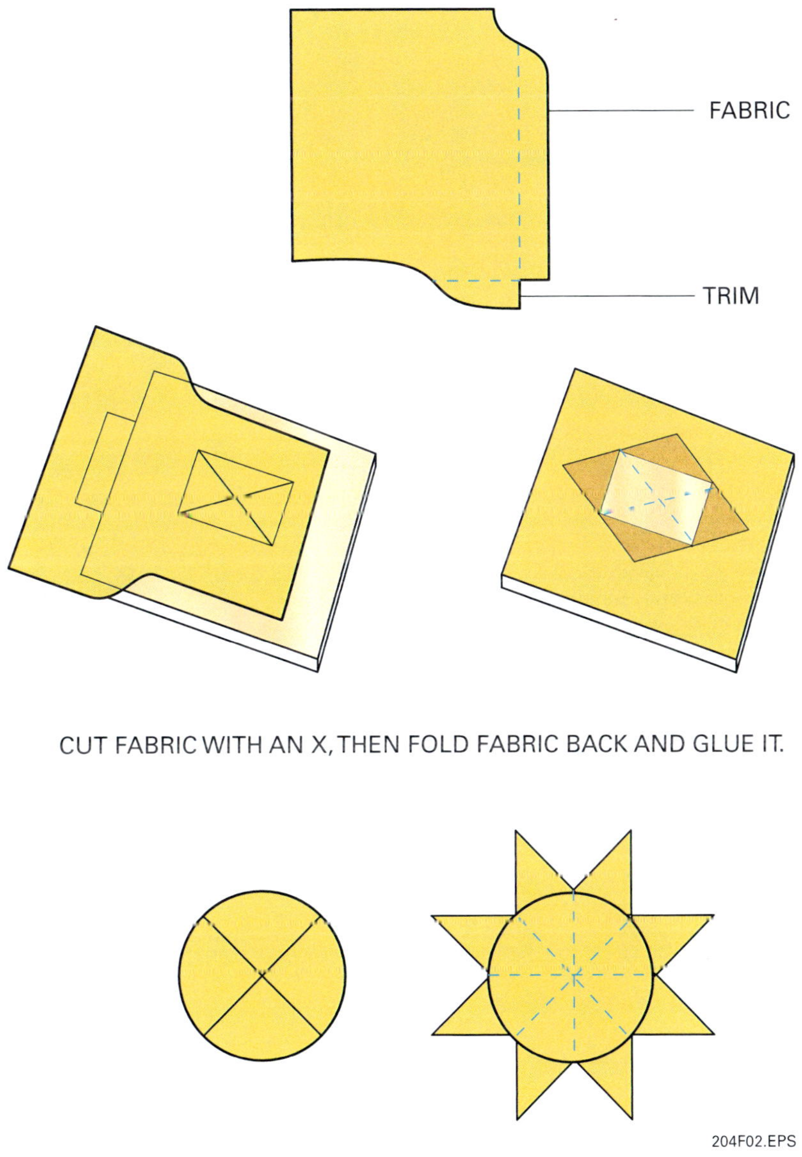

Figure 2 ◆ Trimming a panel.

Fiberglass ceiling panels are typically secured in the same manner as standard suspended ceiling tiles. (See the *Acoustical Ceilings* module for more information.) Fiberglass wall panels can be installed in one or more of the following ways. Always be sure to follow the manufacturer's recommendation.

- *Adhesives* – Adhesives, such as Liquid Nails®, permanently secure ceiling and wall panels. Most adhesives set within 10 to 15 minutes, so you need to work quickly. Apply a bead of adhesive around the perimeter of the panels about 2 inches from the edges. Next, apply adhesive to the center of the panel in the shape of the letter S or W. Carefully set the panel in the desired location, and gently but firmly press it into place. Be sure to align the panel with the chalklines. Small panels can be held in place for a few moments until the bond is secure. Larger panels will probably require some type of support. Impaling clips may be used for both fabric and vinyl covered panels. Nails will work for some fabric-covered ones.

NOTE

Moisture in uncured concrete prevents adhesives from bonding to the surface. To use adhesive on uncured concrete surfaces, install furring strips to the wall, and secure the adhesive to the strips.

WARNING!

Adhesives often contain ingredients that can cause eye, respiratory, and skin irritations. Always check the product's MSDS for appropriate safety precautions. Be sure to use the proper personal protective equipment. Some adhesives out-gas volatile organic compounds (VOCs), so be sure to use them with adequate ventilation.

- *Impaling clips* – During installation, impaling clips (*Figure 3*) help to support fiberglass panels until the adhesive sets. Impaling clips are a rectangular arrangement of sharp pins. The clips are attached to the wall with screws. The pins pierce the back of the fiberglass panel when it is positioned against the wall, holding it in place. The clips are not designed to permanently support the panels. When installing impaling clips, make sure that the screw is installed flush with the wall. Also, be certain to press the panel firmly into place to avoid gaps between it and the wall.

GOING GREEN

VOCs

Many low-VOC adhesives are now available. When given the choice, always select low-VOC products over those that emit high amounts of VOCs.

- *Zee-clips* – Zee-clips are attached to the back of panels, and then the panels are hung on a wall bar that is attached to the wall (*Figure 4*). There are some variations in the design of the bars. Some are attached together to the wall in an unbroken line. Another type of bar is called the wall clip, which is much shorter than the wall bar. The wall clip supports only one zee-clip, so at least two clips must be installed per panel. Another style of bar is the double wall clip. The double wall clip is longer than a regular wall clip, and it supports two zee-clips—one on each of two adjacent panels. Bars must be hung so they are level. Use a laser level and snap a chalkline on the wall at the correct level. Sometimes hook-and-loop fasteners are used with zee-clips for added stability. Always consult the manufacturer's recommendations before using these fasteners.

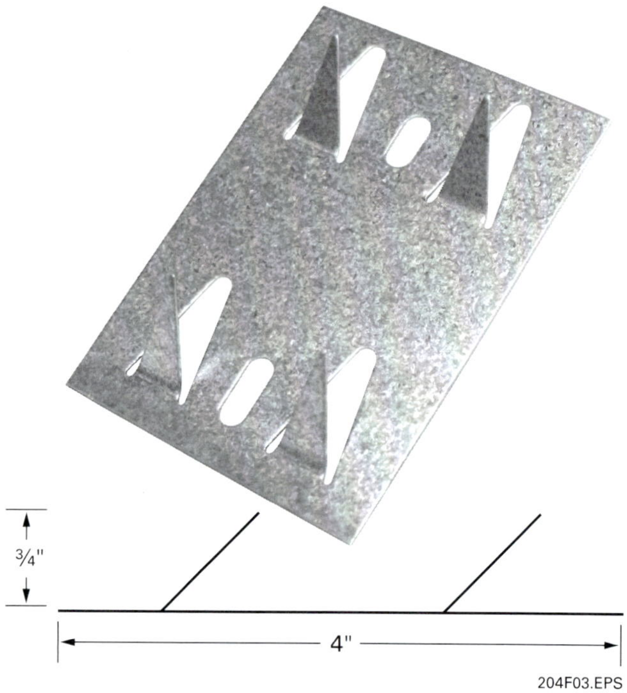

Figure 3 ◆ Impaling clip.

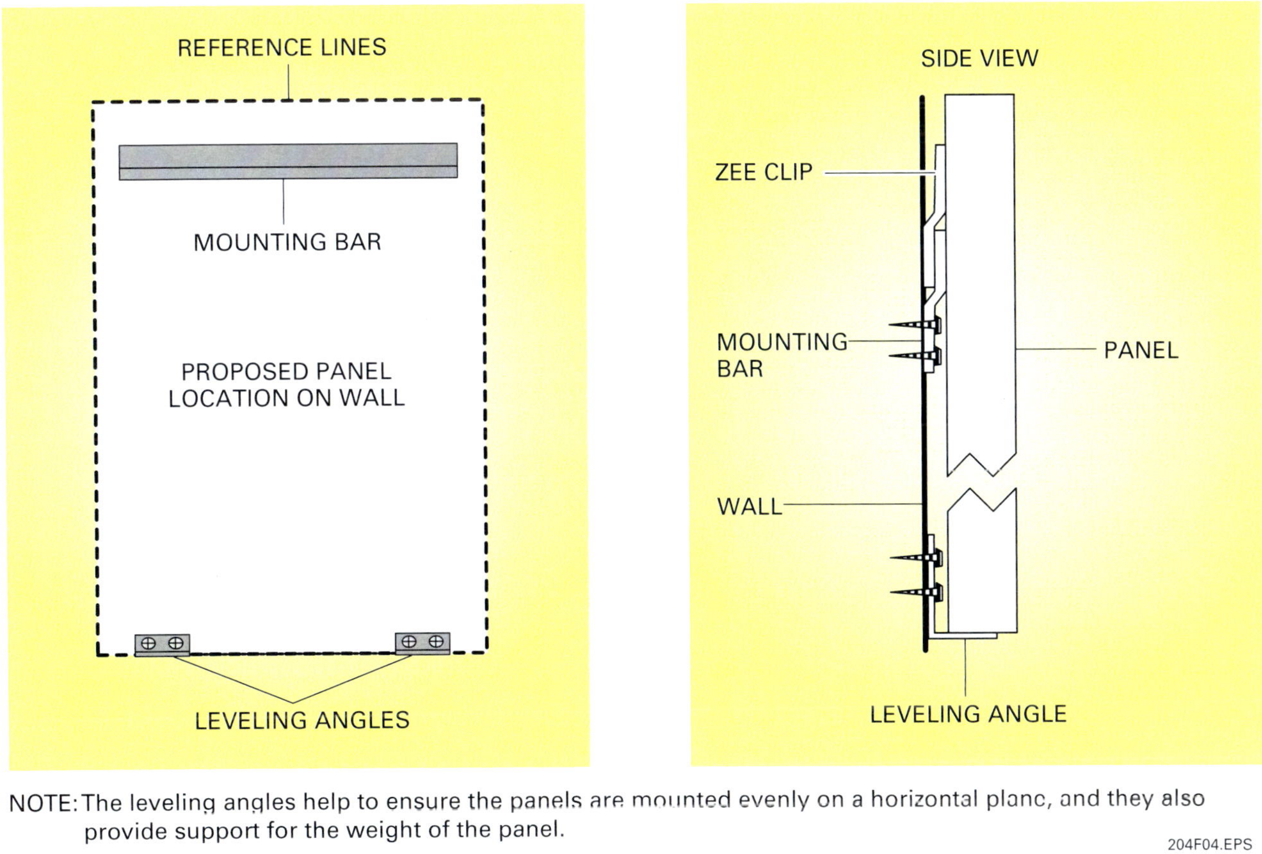

Figure 4 ◆ Zee-clips.

- *VELCRO® hook-and-loop fasteners* – These fasteners are often simply referred to as Velcro. They can be used to secure panels that need to be removable. They can also be used with other types of fasteners to provide additional support.

NOTE

VELCRO® brand hook-and-loop fasteners are available in regular and industrial strengths; never substitute one for the other.

- *Nails and screws* – Nails and screws can be used to secure panels to walls and ceilings. They are not usually the first choice, because they need to be sunk into the panels to hide them. This makes them very hard to remove. When using screws or nails, they must be placed so they are behind the fabric or disguised with a decorative cover. Finishing nails and screws can be placed and countersunk in panels covered with loosely woven fabric, but they can damage vinyl or finely woven fabrics.
- *Battens* – Battens are thin strips of wood cut in symmetrical, trapezoidal shapes (*Figure 5*). One strip is mounted on the wall and another on the back of the panel. The panel is then hung on the batten on the wall. Battens allow panels to be easily removed without damage.

3.0.0 ◆ FABRIC WALLS AND CEILINGS

Fabric walls and ceilings can add a distinctive finish to a commercial area. The fabric can be stretched over some type of framing system and then attached so that fasteners are invisible. This type of installation is also called wall upholstery. Fabric can be glued to the wall just like wallpaper. Fabric wall and ceiling coverings are light in weight and easy to install. When mounted on a framing system, the fabric can be easily removed and replaced. These coverings are often installed with an acoustical material backing to absorb sound.

When the fabric is glued to the wall, surface preparation includes patching any surface defects, and cleaning dust, dirt, grease, loose paint, and old wall coverings from the wall. If necessary, prime or paint the wall in a solid, neutral color and allow it to dry thoroughly before the fabric is applied. This prevents color variations from being seen through the fabric. When

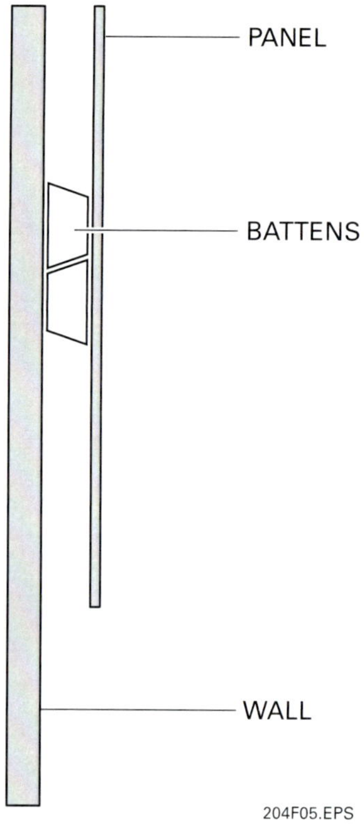

Figure 5 ◆ Battens.

recommended by the manufacturer, snap chalklines that can be used as guides when installing the fabric.

Some wall systems use proprietary framing systems. Be sure to follow the manufacturer's instructions. The following is a general procedure for installing a fabric wall system:

Step 1 Determine the desired position of the frame, and then use a level and chalkline to snap vertical and horizontal lines that will be used as guides during installation.

Step 2 Measure and cut the frames to the desired lengths.

Step 3 Secure the frames to the wall on the lines snapped in Step 1 using the method recommended by the manufacturer.

Step 4 Apply acoustical material to the wall.

Step 5 Attach the fabric to the frame using the tool specified by the manufacturer. Be sure to keep the fabric straight and smooth.

4.0.0 ◆ GLASS FIBER REINFORCED GYPSUM

Glass fiber reinforced gypsum (GFRG or GRG) is a drywall product that has glass fibers mixed in with it. Glass fibers give GFRG strength, so it can be thinner than standard drywall. It is also light in weight and easily molded into various shapes. These traits make GFRG desirable for ornate building features, such as columns, moldings, and arches. In the past, these features were created by skilled craftsmen using costly materials. GFRG's light weight makes it desirable for unique ceiling treatments, such as domed ceilings, light pockets, and vaulted ceilings (*Figure 6*).

Properly primed GFRG can be finished with water-based paints, gold leaf, or stains. It can be finished to look like wood, metal, or marble. In addition, it is fireproof and will not be a source of combustion during a fire. Since GFRG is drywall, it may only be used in areas with low humidity. Glass fiber reinforced concrete (GFRC) must be used in high humidity areas, such as fountains, pools, and other similar locations. For additional information about GFRC, consult the *Exterior Cladding* module.

Many GFRG products are custom-made for a specific location, so they may have unique installation instructions. Off-the-shelf items are typically designed to fit into standard framing. The following are general guidelines about GFRG products:

- Carefully inspect shipping containers when they arrive at the job site. Note any damage on the shipping invoice; if necessary, contact the manufacturer for instructions.
- Store products in a safe, dry location until they are installed. Don't unpack containers unless instructed to do so by the manufacturer. An exception to this occurs when the purchase agreement states that products must be rejected on delivery. When this is the case, unpack the product, carefully inspect it, and then repack it for storage.
- Never install damaged products.
- Install products according to the manufacturer's instructions, using the recommended hardware and adhesives. Seal all joints with the recommended sealants.
- Keep a variety of shims on hand to help align the pieces as they are installed.
- When installing ceilings, use wire hangers and straps, support wires, and anchoring devic-

204F06.EPS

Figure 6 ◆ Examples of glass fiber reinforced products.

es, just as with standard ceiling installations. Seismically active areas will require additional supports.

- GFRG joints are finished using standard drywall procedures and supplies, such as tape and joint compound.
- GFRG products must be primed as directed by the manufacturer and then finished as stated in the project specifications. GFRG products are often painted in marble or gilded faux finishes.
- Remember that GFRG is drywall and is subject to expansion and contraction with changes in humidity. When expansion joints are required, install them exactly as specified in the manufacturer's instructions.
- Consult shop drawings for specific panel arrangement and layout.

5.0.0 ◆ WOOD WALL AND CEILING SYSTEMS

Wood wall and ceiling systems are engineered for easy installation and maintenance. Some of these systems are designed to be installed on both walls and ceilings. Most of these systems use wood **veneer** instead of solid wood for the finished surface. Veneer is a thin layer of wood that is glued to an engineered material. Veneers minimize the amount of wood used in the product, decreasing costs.

As with most commercial wall and ceiling systems, wood systems (*Figure 7*) are typically designed to improve the acoustical properties of an area. However, a smooth, flat surface can reflect sound waves in such a way that an echo is produced. To prevent this, the surface of these wood products is usually slotted, drilled with small holes, or grooved to change the way sound waves are reflected. In addition, these systems are typically backed with acoustical batting.

Wood panels are usually attached to walls with zee-clips, screws, or a combination of the two. Some wood panels are joined together by interlocking channels, such as tongue and groove joints (*Figure 8*). Wood can swell and shrink with increases and decreases in humidity. Because of this, it is very important to expose the product to the climate of the area where it will be installed. This is done by storing the containers holding the product for a specified period of time at the location where they will be installed. This allows the product to adjust to the temperature and humidity of the area. If the containers are wrapped in plastic film, it may need to be removed prior to storage. Check the manufacturer's instructions for detailed information.

Linear ceilings are lightweight strips of wood, or slats. The slats are clipped to a frame that is suspended from the ceiling by wires, rods, or some other type of hanging device. These systems may be fitted with lighting fixtures and heating/air conditioning vents. Some ceilings are backed with a sound absorbing material for noise reduction (*Figure 9*).

One disadvantage of wooden slats is that they can warp and twist. Warped or twisted slats cannot be installed because they will not seat properly. Before installation, each slat must be checked to make sure that it is straight. This inspection can slow down the installation of a ceiling. One option is to separate unsuitable slats before starting the installation work. If this is done, the slats must be handled carefully to avoid damaging the finish. Return twisted or warped slats to the manufacturer for replacement.

204F07.EPS

Figure 7 ◆ Wood wall system.

204F08.EPS

Figure 8 ◆ Tongue and groove joint.

204F09.EPS

Figure 9 ◆ Wood linear ceilings.

As with all wall and ceiling systems, be sure to follow the manufacturer's instructions during installation. The following procedure may be used as a general installation guide:

Step 1 Study the reflected ceiling plan (RCP), the manufacturer's drawings, and/or the room finish schedules to find out the proper direction for orienting the wooden slats during installation, and to determine the ceiling elevation. When the area where the ceiling will be installed is multi-level, be sure to determine the ceiling elevation for each level. This is of special importance when the area has ramps and stairs.

Step 2 Using a laser level and a chalkline, snap a line at the desired ceiling elevation.

Step 3 Determine the level at which the ceiling perimeter trim should be installed. Depending on the system being used, you may need to add the thickness of the ceiling slat to the desired ceiling elevation to find out the perimeter trim elevation. For example, if the ceiling slat is ¾" thick and the ceiling elevation is 11'-9", then the ceiling trim should be installed at 11'-9¾". Check the manufacturer's instructions for more details.

Step 4 Install the ceiling perimeter trim according to the job specifications.

Step 5 Snap chalklines at the desired location of the cliprails. In the example shown in *Figure 10*, the cliprails are installed 24" OC with a 4" allowance at the edges.

Step 6 Attach hanger clips to the ceiling in accordance with local code, and then fasten the hanger wires to the clips.

Step 7 Secure the cliprails to the hanger wire, ensuring that all cliprails are level (*Figure 11*). In this case, 12-gauge hanger wire is used. Mate the cliprail pieces together as necessary to reach the desired length.

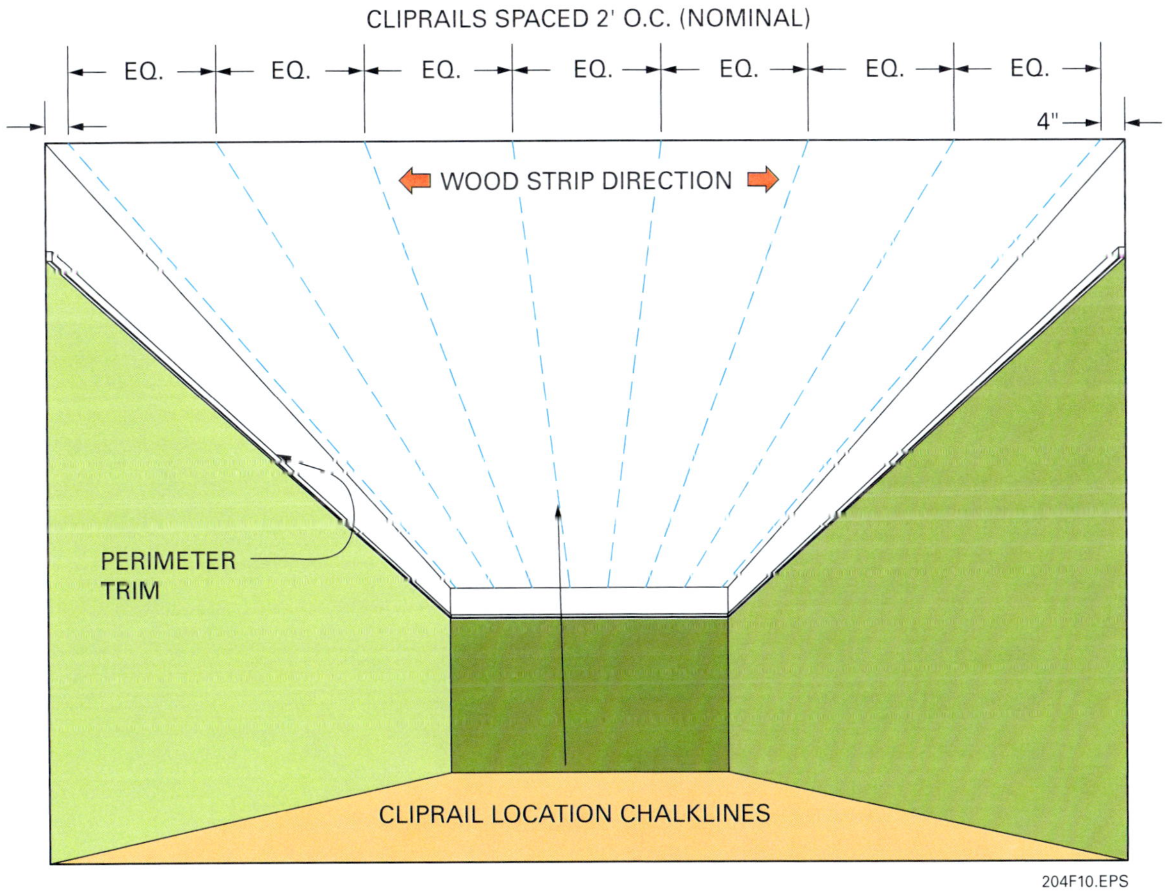

Figure 10 ♦ Snapping chalklines for cliprail placement.

Step 8 Examine each slat before you install it, and make sure that it is straight. Return twisted or warped slats to the manufacturer for replacement. Never install defective slats.

Step 9 Starting at one end, snap the wooden slats into the connectors on the cliprail. The manufacturer may require the use of an installation tool. Some designs overlap, such as the one shown in *Figure 12*.

Step 10 If recommended by the manufacturer, secure the slat to the rail with screws placed at the recommended intervals (*Figure 13*).

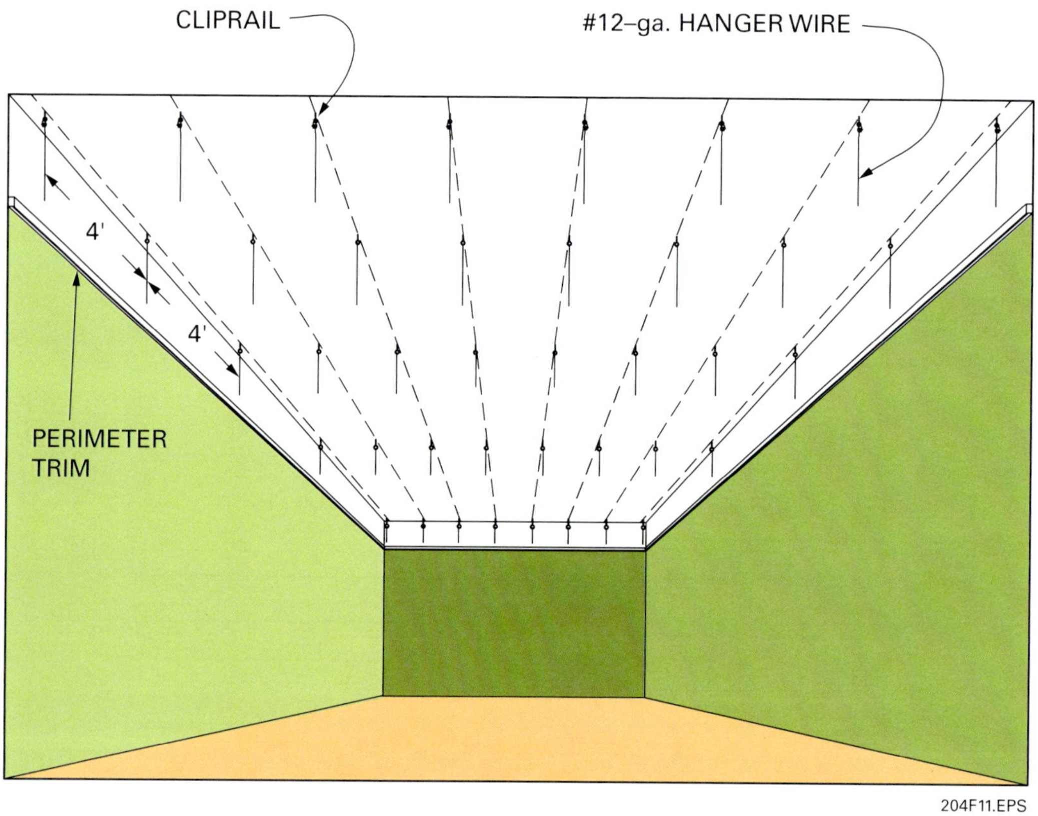

Figure 11 ◆ Suspending cliprails from hanger wires.

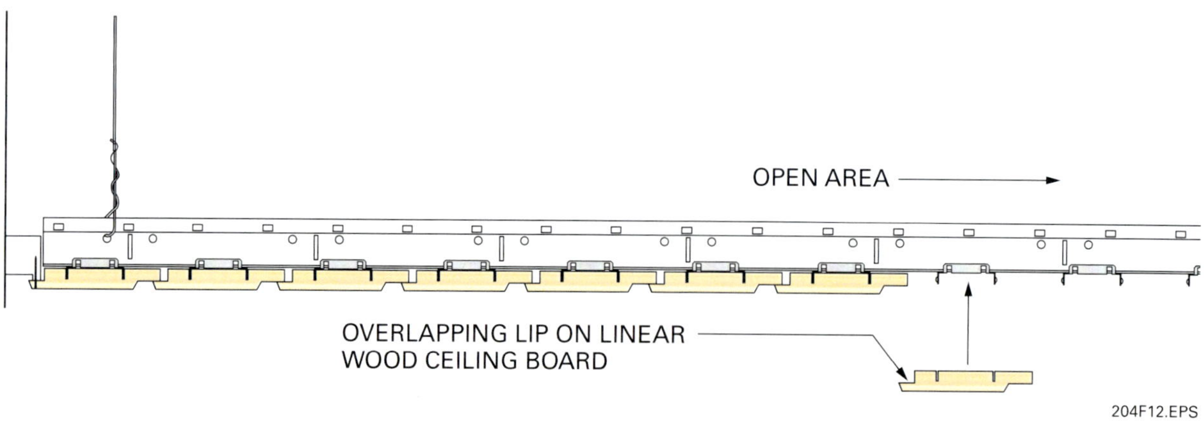

Figure 12 ◆ Installing slats onto the cliprail.

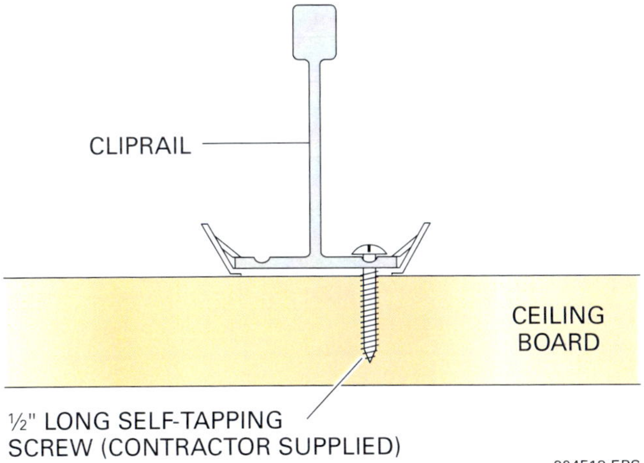

Figure 13 ◆ Securing the slat with screws.

6.0.0 ◆ CEILING TRIM SYSTEMS

Ceiling trim systems are used on suspended ceilings that are not attached to the walls. The trim is attached to the edge of the ceiling grids and panels to give them a finished look. In trim systems, suspended ceiling panels can be used to make unique ceiling features, such as ceiling islands (*Figure 14*). These islands are often used to diffuse overhead lighting, control acoustics, and add visual interest to an area.

Notice in *Figure 14A* that the ceiling above the island is covered with the same type of tiles as the island. In addition, the lighting fixture is located between the upper ceiling and the island. This photograph is an example of a suspended ceiling (the island) below a suspended ceiling.

Trim systems come in many shapes and colors, and they can be used with a variety of ceiling systems. Always check the manufacturer's recommendations for ceiling panel and trim compatibility before you begin a job.

Keep the following points in mind when you are installing ceiling trim systems:

- Study the RCP or the manufacturer's shop drawings for information about the suspended ceiling design and arrangement.
- All suspended ceiling hardware must be attached to the upper deck of the area. This is especially important when an island is installed below a suspended ceiling. Never secure an island's framework to a suspended ceiling unless it is recommended by the manufacturer.
- Ceiling trim is applied to the edges of suspended ceilings. The ceilings themselves must be installed in accordance with the appropriate local code, including seismic factors. When installing ceiling trim systems, use wire hangers and straps, support wires, and anchoring devices, just as with standard ceiling installations. Note that some trim systems require that the trim is secured to the upper deck with wire hangers, straps, or other anchoring devices. Be sure to check the manufacturer's instructions for details.

Figure 14 ◆ Compasso™ suspension trim.

- Never cut any ceiling hanger devices in order to install a trim system. Doing so will weaken the structural integrity of the ceiling system.
- When ceiling islands are installed below a suspended ceiling, some equipment, such as fire sprinklers, must be installed through the island. Other equipment, such as lighting fixtures, may be installed either through the island or above it. Always check the RCP, project specifications, and construction plan detail drawings (if available). When fixtures are installed above islands, check to see if the upper side of the island's tiles will require some type of facing material. For example, when lighting is installed above an island, reflective material is often placed on the upper side of the tiles.
- Always handle the trim carefully to avoid scratching the finish.
- Never mix pieces from different trim systems.
- Wash your hands before handling the trim pieces to avoid soiling the finish.

Trim system components are designed to be versatile and to minimize the number of cuts that need to be made in the field. Trim pieces may be straight or curved and may be mixed to create the desired shape. The Gordon VANTAGE POINT™ Contura perimeter trim shown in *Figure 15* uses straight trim pieces and 90-degree corners. For the most part, this type of design is a matter of correct assembly using splice plates, T-Barr connector clips, and corner hardware supplied by the manufacturer. The screws on these connectors need to be tightened until they are snug, but they should not be overtightened, which can warp the trim.

The ceiling islands shown in *Figure 14* use curved pieces of Compasso™ suspension trim. This type of product allows the trim to be shaped into many attractive designs. *Figure 16* also shows examples of straight pieces of this trim. The curved pieces are a bit harder to install than the straight ones. The procedure for placing a trim system will be somewhat different for each installation. The following steps may be used as a general installation guide:

Step 1 Install the ceiling grid to which the trim system will be attached. Make sure that the grid is larger than the island.

Step 2 Plan the trim installation by studying the RCP.

Step 3 Place the trim on top of the suspended grid.

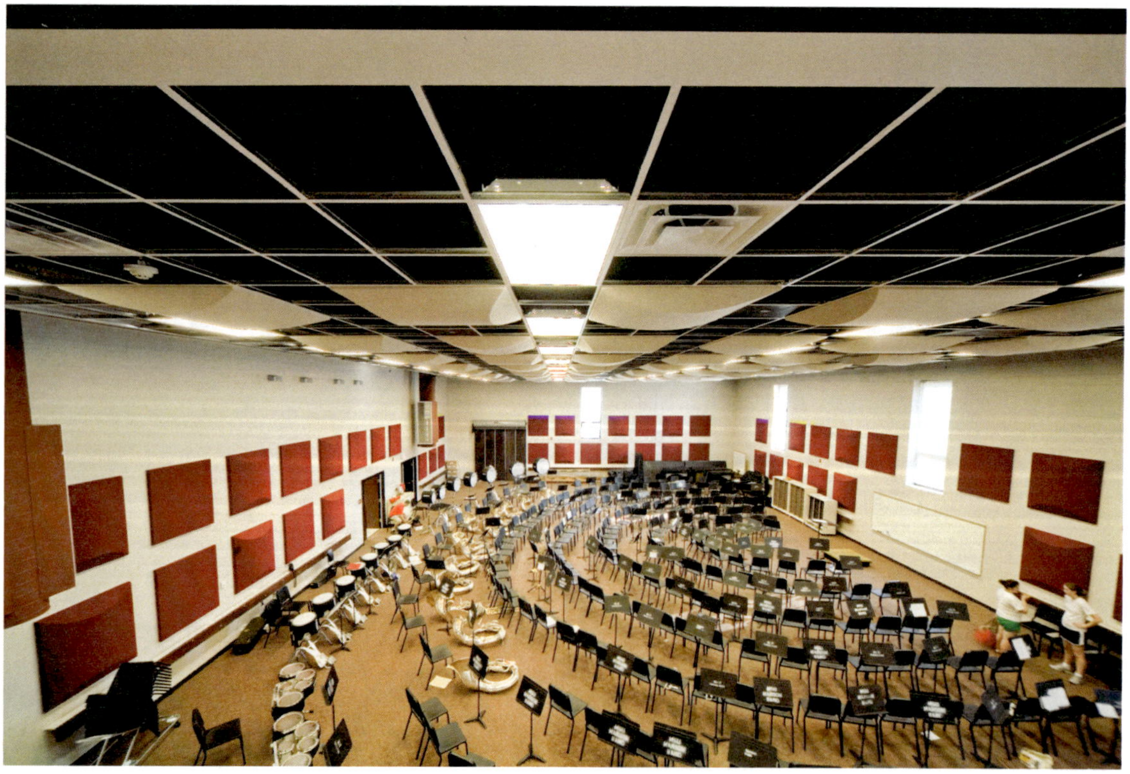

204F15.EPS

Figure 15 ◆ Acoustic ceiling islands.

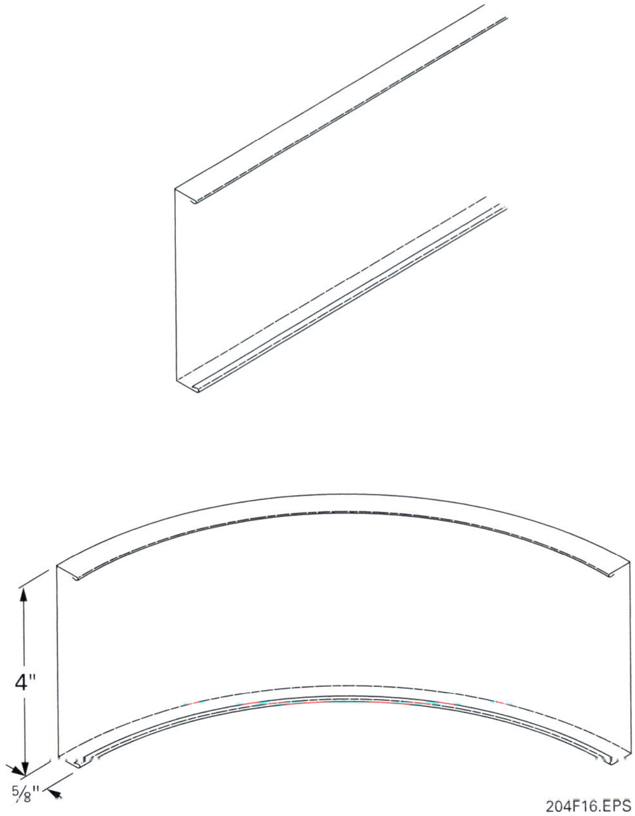

Figure 16 ◆ Compasso™ suspension trim (straight and curved).

Step 4 Using binder clips, fasten the trim pieces to the suspended grid in the desired shape. If necessary, place a paper towel or clean cloth over the trim to protect the finish before applying the clips.

Step 5 After all trim pieces have been laid out, use the appropriate splice connectors to join the trim pieces and form the desired shape (*Figure 17*). You will need to remove the binder clips to do this.

Step 6 Position the trim as desired on the grid, and square the suspension grid as needed. Then reattach the binder clips.

Step 7 Mark the tees wherever they intersect with the trim.

Step 8 Using aviation snips, trim the tees as necessary.

Step 9 Fasten an attachment clip to the trim, and then insert the tee into the clip. Use a screw to attach the clip to the tee. Compasso™ trim pieces in widths of 10 inches and 12 inches require that diagonal bracing is used with attachment clips.

Step 10 Cut the ceiling panels to fit, and then install them into the grid.

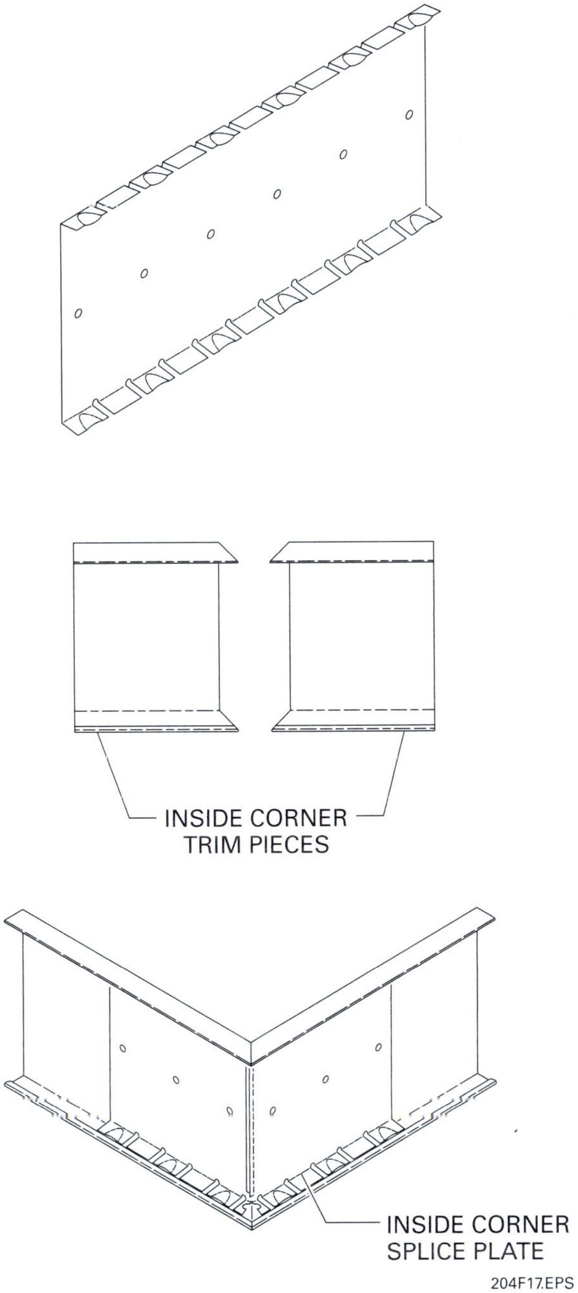

Figure 17 ◆ Compasso™ suspension trim (splice pieces).

7.0.0 ◆ TECTUM™ PANELS

Tectum™ products are made with wood fibers that are sprayed with a proprietary formula of concrete material and then molded into panels (*Figure 18*). These panels have a rough, irregular, fibrous surface, which gives it good acoustical properties. The material is durable and can withstand some abuse. It is often used on walls and ceilings in schools and gymnasiums. Because the wood fibers are coated with concrete, Tectum™ panels are resistant to humidity and are often used near swimming pools. The panels cannot resist standing water, however.

204F18.EPS

Figure 18 ◆ Tectum™ panels.

Like standard acoustical ceiling panels, Tectum™ ceiling panels are installed using lay-in, concealed tee, or profiled grid systems. Wall panels can be installed in the same manner as fabric- and vinyl-covered acoustical panels, except when used in high humidity areas. In these areas, the panels are usually installed on furring strips instead of flush with the wall. This allows the air to circulate behind the panel.

Tectum™ products may be used with their natural finish, may be prefinished, or they may be field painted, but painting them requires some caution. The wood fibers in the panels form random spaces and holes in the panels and on the surface. The spaces and holes on the surface give the material desirable acoustical properties. Paint can fill the spaces and change the acoustical properties of the material. Most Tectum™ products can be painted with up to six layers of paint before their acoustical properties decline. When spray painting Tectum™ products, be sure to use only enough paint to cover the fibers.

Tectum™ roofing panels are often used for the upper deck of sports facilities (*Figure 19*). The panel shown in *Figure 20* combines an interior finish surface (Tectum™ panel), insulation (expanded polystyrene), and OSB sheathing.

204F19.EPS

Figure 19 ◆ Tectum™ roof deck.

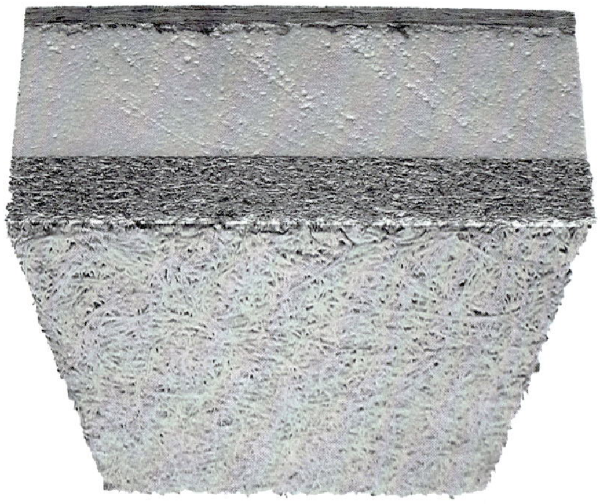

204F20.EPS

Figure 20 ◆ Tectum™ E composite panel.

GOING GREEN

Tectum™ Products

An added benefit of Tectum™ products is that they are biodegradable. These durable panels also contain no toxic binders, asbestos, or formaldehyde.

8.0.0 ◆ OTHER FINISH SYSTEMS

There are many types of finish systems available. This section briefly discusses ceiling domes, metal columns, and reveal trim.

8.1.0 Domed and Vaulted Ceilings

Some companies, such as USG Corporation, sponsor computer software programs that allow contractors to easily design custom domes (*Figure 21*) and barrel vaulted ceilings (*Figure 22*). USG Corporation's software tool lets contractors fill in key information about the desired installation, and then it automatically designs the feature. The tool also creates customized installation instructions.

The domes and vaults are created from components purchased from the manufacturer (*Figure 23*). The components are used to build a framework grid (*Figure 24*). The feature is hung into place and sheathed with drywall board and/or plaster compound. The surface can be finished in the same manner as any drywall product.

204F22.EPS

Figure 22 ◆ Barrel vaulted ceiling.

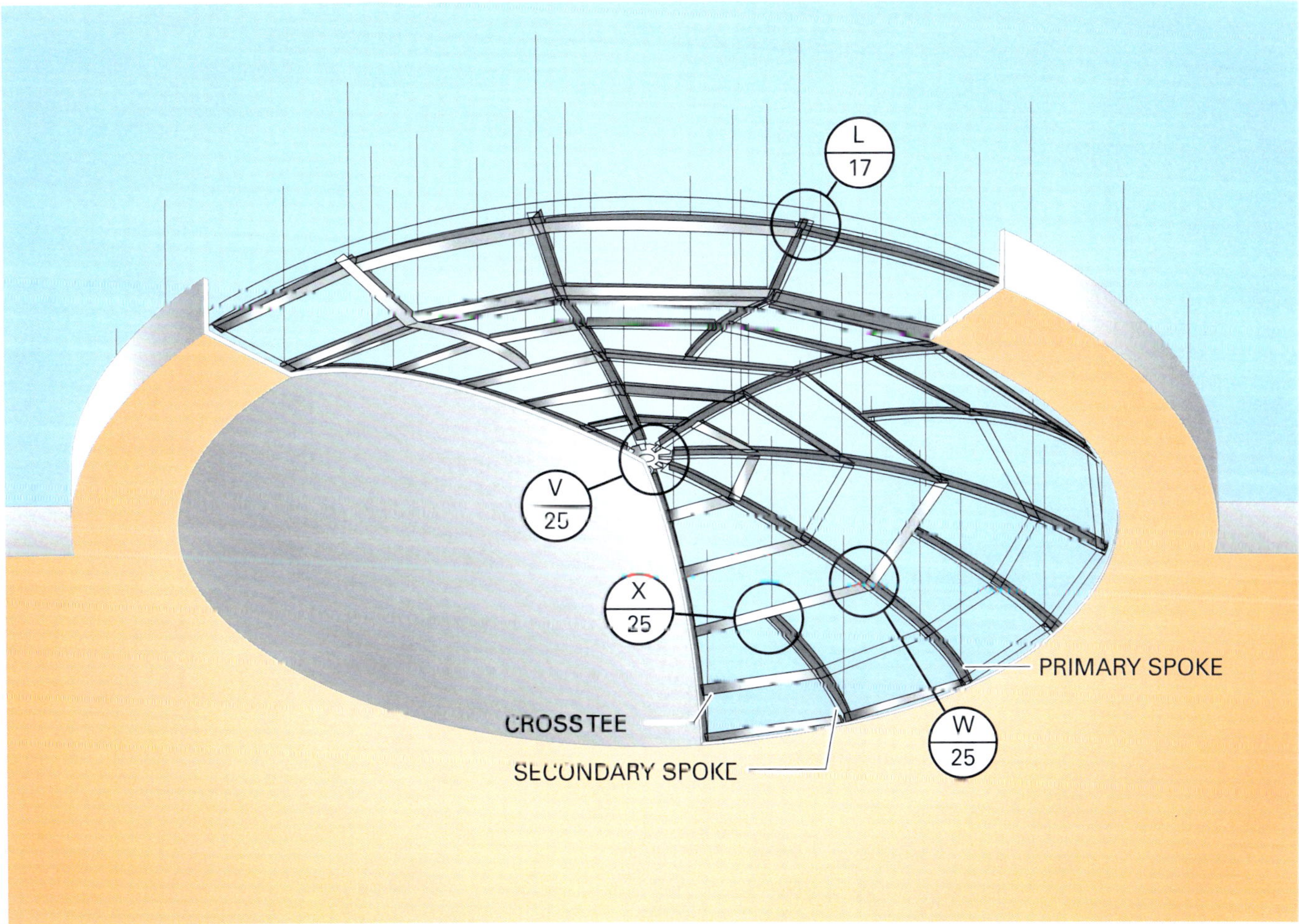

204F21.EPS

Figure 21 ◆ Dome feature.

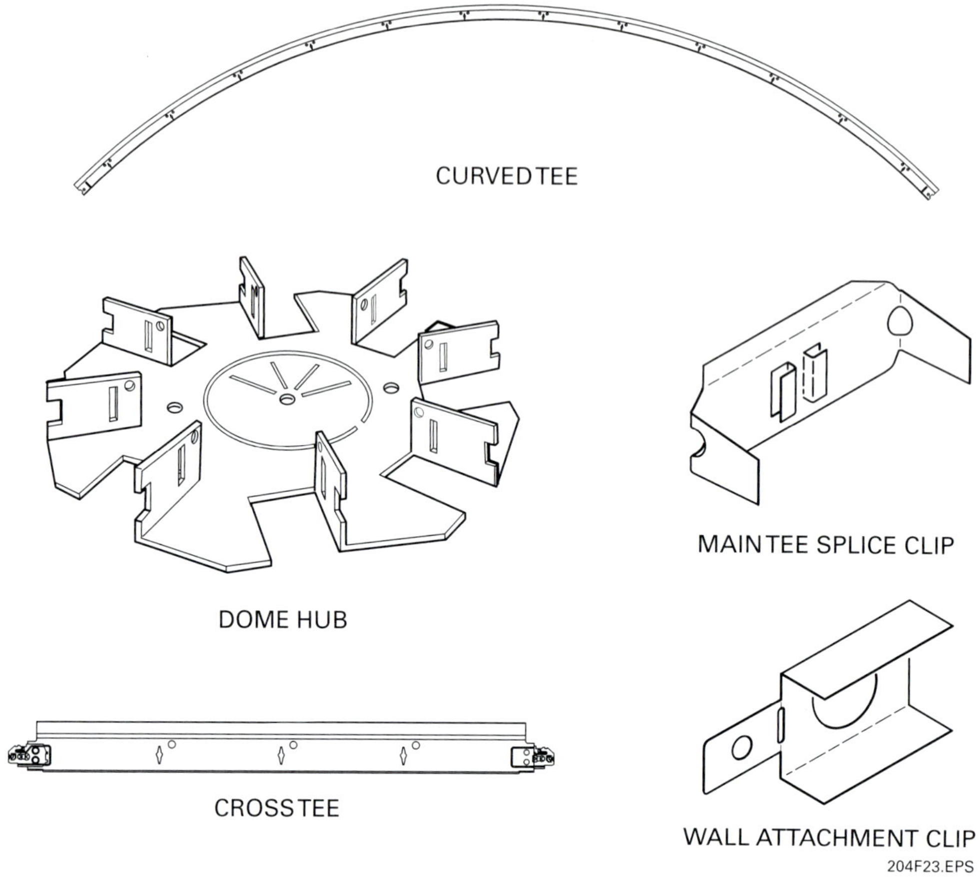

Figure 23 ◆ Dome components.

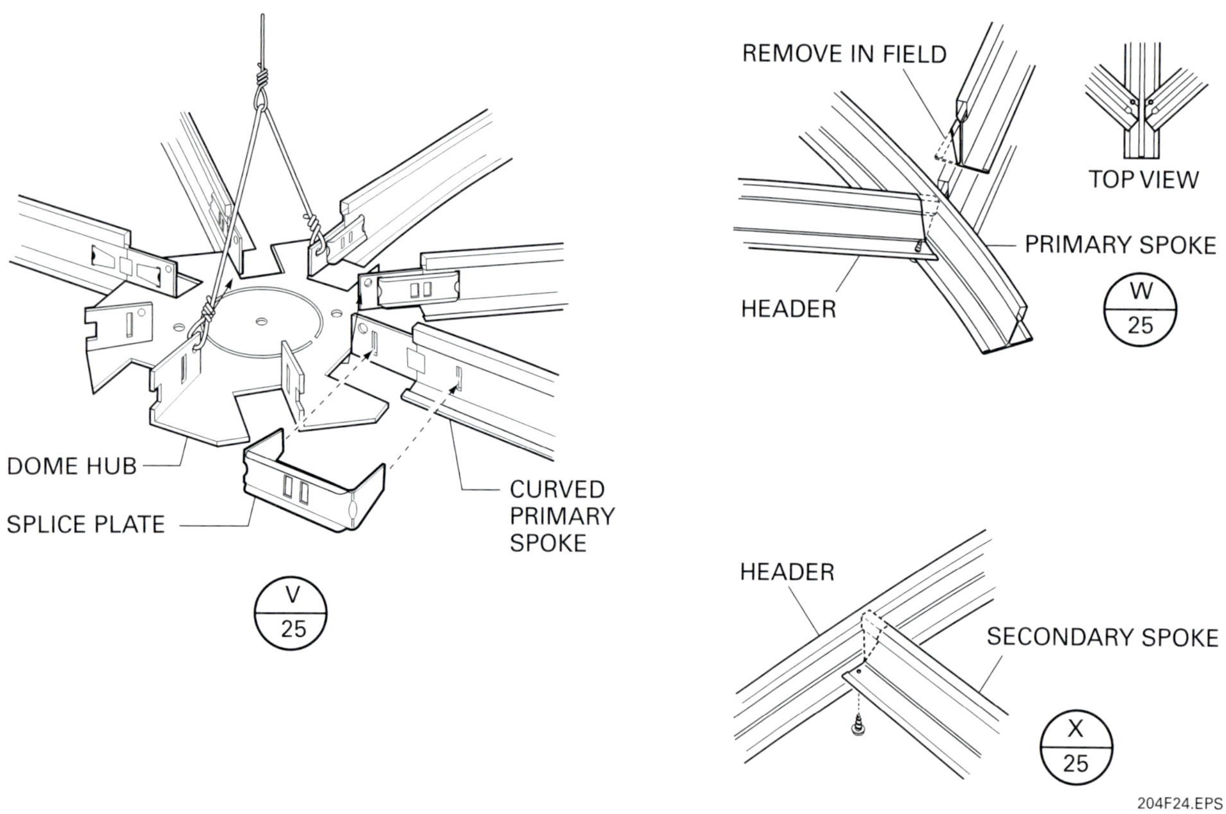

Figure 24 ◆ Dome framework grid.

8.2.0 Metal Columns

Commercial structures are built to be solid and durable. Sometimes the placement of structural components makes them appear unsightly. Many finish systems are designed to disguise components. The metal columns shown in *Figure 25* are used to cover structural members such as I-beams. These covers can easily be attached using one of several different methods.

Metal column covers are available in modern and classical styles. The covers are designed strictly for looks and should never be used to carry a load. When installing the covers, handle them with care to avoid damaging the finish.

8.3.0 Reveal Trim

Reveal trim is used to finish the joints of drywall and acoustical tiles by creating a gap between the panels (*Figure 26*). This gap provides a quick and easy way to finish panel edges and create a visual feature. Reveal trim can be used on both ceilings and walls. It is often used to create a smooth transition between the wall and the ceiling. On walls, reveal trim may run horizontally or vertically.

9.0.0 ♦ SAFETY

Since most finish work is performed above ground level, it is important to remember that

204F25.EPS

Figure 25 ♦ Metal column covers.

204F26.EPS

Figure 26 ◆ Reveal trim.

safety is a priority. Exercise caution when handling sharp tools and metal components. Doing overhead work often requires the use of scaffolding or ladders to reach the work area. Working aboveground even a few feet adds an element of danger to the job. Use the correct height of equipment for each job, and resist the urge to over-reach for your equipment. Be sure that all equipment is in good working order. Any defective items must be tagged immediately so that they will not be used.

Scaffolding must be inspected by a competent person before it is used on each shift, and whenever it is moved or modified in any way. Some scaffolding must be designed by an engineer, so know the practices at your work site. If you are unsure of the site requirements, ask your supervisor before you begin work.

When working above ground, wear the appropriate personal protective equipment, maintain control of materials and tools at all times, and stay alert for personnel working on the ground below. If you are working on the ground when work is being done overhead, wear the appropriate personal protective equipment and stay alert for falling debris and tools.

When using any product, first check its MSDS. Use the recommended safety precautions. Never smoke where adhesives, paints, solvents, or other chemicals are being used. Wash your hands before eating, before and after using the rest room, and when you are leaving the work site.

Review Questions

1. The best place to store fabric-covered fiberglass wall panels is _____.
   a. leaning against a wall in a storage room inside the building
   b. carefully covered anyplace outside the building
   c. leaning against a wall anyplace inside the building
   d. carefully covered in a storage room inside the building

2. After trimming the fabric on a fabric covered panel, Liquid Nails® adhesive should be used to reattach the fabric to the panel.
   a. True
   b. False

3. When a square hole for a light fixture must be cut in a fabric-covered fiberglass panel, the type of cut to make in the fabric before folding it back is _____.
   a. X-shaped
   b. Y-shaped
   c. W-shaped
   d. V-shaped

4. Adhesives that are used to secure fiberglass panels to walls set in about _____.
   a. 10 to 15 minutes
   b. 30 to 60 minutes
   c. 1 to 4 hours
   d. 24 hours

5. The screws that attach impaling clips to the wall should _____.
   a. protrude slightly from the wall
   b. be mounted flush with the wall
   c. be coated with adhesive
   d. be disguised with a decorative cover

6. Screws can be used to mount fiberglass panels that are covered with _____.
   a. loosely woven vinyl
   b. tightly woven fabric
   c. tightly woven vinyl
   d. loosely woven fabric

7. Before installing the framework for upholstered walls, you should _____.
   a. scrape old paint from the wall
   b. tape the fabric to the wall to make sure that it fits properly
   c. thoroughly wash the walls
   d. snap horizontal and vertical chalklines to be used as guides

8. The glass fibers in glass fiber reinforced gypsum (GFRG) prevent the expansion and contraction that can occur with changes in humidity.
   a. True
   b. False

9. To decrease costs, most wood ceilings and wall systems use _____.
   a. slats
   b. planks
   c. veneers
   d. panels

10. The acoustical properties of wood wall panels are improved by _____.
    a. drilling holes or cutting slots in the panels
    b. staining them
    c. installing panels so air circulates behind them
    d. painting them

11. When installing trim systems on a suspended ceiling island, _____.
    a. it is okay to cut a hanger device, if necessary
    b. secure the island hangers to the suspended ceiling above it
    c. install the fire sprinklers through the island
    d. be careful not to splash the panels when you paint the trim

Review Questions

12. To protect the finish on the pieces of a trim system, you should _____.
    a. dust your hands with talcum powder before handling trim
    b. keep the trim covered during installation
    c. wash your hands before handling trim
    d. firmly place binder clips so the trim will not slip during installation

13. When installing trim systems on a suspended ceiling island, trim the tees with _____.
    a. a circular saw
    b. power scissors
    c. GFRG
    d. aviation snips

14. Tectum™ products are made with _____.
    a. wood fibers sprayed with concrete material
    b. glass fibers mixed with adhesives
    c. fiberglass
    d. fabric fibers

15. Tectum™ panels in a pool area are usually installed _____.
    a. with adhesive
    b. away from the pool
    c. so that air circulates behind them
    d. where they cannot be splashed with water

Summary

Finish work is what gives modern commercial buildings their distinctive appearance. Finish systems are available in many materials and designs. Some give an area a sleek modern look, while others duplicate popular designs from the past. The finishes used on commercial interiors must look attractive and be easy to install and maintain. In addition, almost all finish systems used today are specifically designed to improve the acoustical properties of an area.

Notes

Trade Term Introduced in This Module

Veneer: A thin sheet of material, such as wood, that is glued to some type of fabricated material.

Additional Resources and References

Additional Resources

This module is intended to be a thorough resource for task training. The following reference work is suggested for further study. This is optional material for continued education rather than for task training.

Drywall: Professional Techniques for Great Results, 2002. Myron R. Ferguson. Newton, CT: Taunton Press.

Figure Credits

Auralex Acoustics, Inc., 204F01, 204F03

Images used with permission from RWM Incorporated (www.rwm-inc.com). All rights reserved, 204F06

Fry Reglet Architectural Metals, 204F07, 204F25

9Wood, Inc., 204F09

USG Corporation, 204F14, 204F16, 204F17, 204F21–204F24

Gordon Inc./www.gordonceilings.com, 204F15, 204F26

Tectum Inc., 204F18–204F20

NCCER CURRICULA — USER UPDATE

NCCER makes every effort to keep its textbooks up-to-date and free of technical errors. We appreciate your help in this process. If you find an error, a typographical mistake, or an inaccuracy in NCCER's curricula, please fill out this form (or a photocopy), or complete the online form at **www.nccer.org/olf**. Be sure to include the exact module ID number, page number, a detailed description, and your recommended correction. Your input will be brought to the attention of the Authoring Team. Thank you for your assistance.

Instructors – If you have an idea for improving this textbook, or have found that additional materials were necessary to teach this module effectively, please let us know so that we may present your suggestions to the Authoring Team.

NCCER Product Development and Revision

13614 Progress Blvd., Alachua, FL 32615

Email: curriculum@nccer.org

Online: www.nccer.org/olf

❑ Trainee Guide ❑ Lesson Plans ❑ Exam ❑ PowerPoints Other ________________

Craft / Level: Copyright Date:

Module ID Number / Title:

Section Number(s):

Description:

Recommended Correction:

Your Name:

Address:

Email: Phone:

45205-09

Exterior Cladding

45205-09
Exterior Cladding

Topics to be presented in this module include:

Overview

Exterior cladding places the finishing touches on a building, and it protects a building's structural members from the damaging effects of weather. This module explains the similarities and differences between commonly used exterior claddings. In addition, it covers the proper installation procedures to use to ensure that cladding protects the structural integrity of a building.

Objectives

When you have completed this module, you will be able to do the following:

1. Distinguish between extruded and expanded foam insulations.
2. Identify trims used in exterior insulation and finish systems (EIFS) and stucco, and state their uses.
3. Distinguish between traditional and water management EIFS.
4. Distinguish between traditional hard-coat plaster and synthetic finishes.
5. Describe how to install synthetic veneer stone.
6. Describe building features commonly created with glass fiber reinforced concrete (GFRC).

Trade Terms

Control joint
Elastomeric
Flashing
Permeable
Starter strip
Weep screed

Prerequisites

Before you begin this module, it is recommended that you successfully complete *Core Curriculum*; *Drywall Level One*; and *Drywall Level Two*, Modules 45201-09 through 45204-09.

This course map shows all of the modules in the second level of the *Drywall* curriculum. The suggested training order begins at the bottom and proceeds up. Skill levels increase as you advance on the course map. The local Training Program Sponsor may adjust the training order.

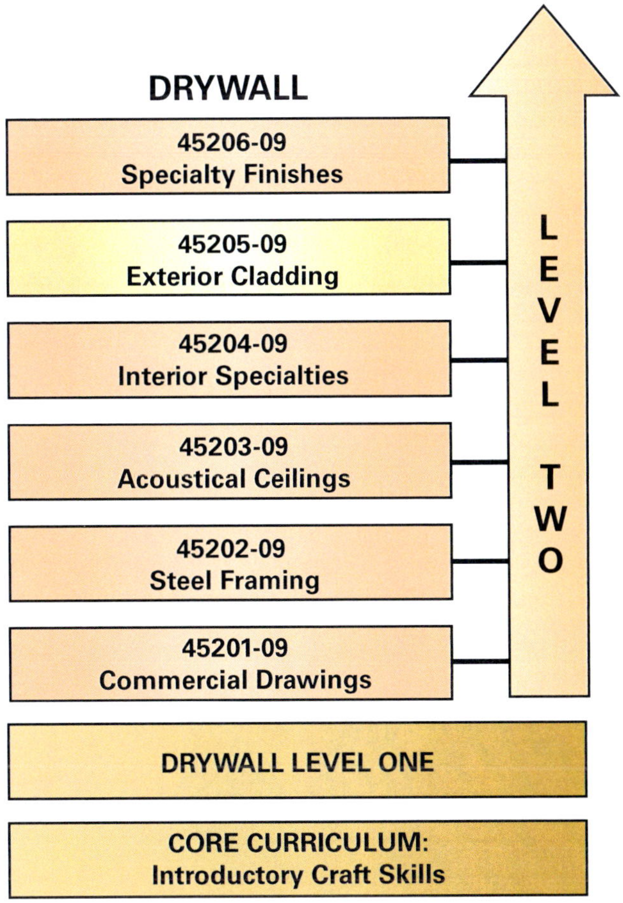

205CMAP.EPS

1.0.0 ◆ INTRODUCTION

Exterior cladding improves a building's appearance, but that is not its primary purpose. One of the main functions of exterior cladding is to protect buildings from the harsh and damaging effects of weather. Another is to help provide the necessary air, moisture, and temperature barrier between the building's interior and exterior. For cladding to do all of this, it must be installed correctly. Improperly installed cladding can make the building's occupants uncomfortable due to drafts and temperature changes. It can also seriously damage the building's structural soundness.

This module provides information about some of the common types of exterior claddings used today, including their similarities and differences. In addition, you will learn about proper installation techniques. The claddings covered in this module include the following:

- Stucco
- Exterior insulation and finish systems (EIFS)
- Synthetic veneer stone
- Panelized cladding
- Glass fiber reinforced concrete (GFRC)

2.0.0 ◆ AIR INFILTRATION

The exterior sheathing of a structure is usually covered to prevent the infiltration of outside air into the structure caused by wind pressure. The covering also needs to be properly designed and installed in order to prevent the introduction of moisture into building cavities, which can result in the growth of mold. To achieve maximum energy efficiency in a structure, air infiltration must be strictly controlled. Traditionally, structures have been covered with water-resistant building paper. This paper helps prevent water which may have leaked through the primary barrier (siding) from reaching the structural sheathing or other components of the structure. The paper is water **permeable** so that moisture inside the walls can pass through and evaporate. To some extent, the paper reduces air infiltration of the structure, especially when board sheathing is used.

House wraps or building wraps are also used in place of building paper. These products, under brand names such as Tyvek®, HardieWrap™, or ProWrap®, are easier to apply and perform the same functions as building paper (*Figure 1*). When properly applied and sealed, the wraps provide a nearly airtight structure no matter what sheathing material is used. Most versions of these wraps are an excellent secondary barrier under all siding, including stucco and EIFS.

Figure 1 ◆ Building wrap.

Building Wraps

GOING GREEN

While not always required by the local code, a properly applied wrap gives a building a weather-resistant seal that will block hot and cold air and help lower heating and cooling costs. For building wrap to be effective, all gaps must be secured properly with tape approved by the manufacturer. Gaps include spaces around doors and windows, as well as any tears in the wrap.

Newer products use spray-on or trowel-on barrier membrane systems. The systems include the membrane and the associated **flashing**. Common manufacturers include Sto, W.R. Grace, and Henry Company. The advantage of these systems is that they provide a seamless **elastomeric** barrier and are therefore somewhat self healing.

Many wraps are made of spun, high-density polyethylene fibers randomly bonded into a very tough, durable sheet material. These wraps are usually available in several types and weights for residential and light commercial use. Special versions have vertical water channels permanently pressed into the material. These channels help to drain water that seeps behind the cladding system. The material is usually furnished in rolls in various widths from 18 inches to 10 feet, and in lengths from 100 to 200 feet.

There are various ways to secure the wrap. Common methods include nails with large heads, nails or screws with plastic washers (*Figure 2*), or 1-inch wide staples, depending on the manufacturer's instructions. Screws and washers are used for steel construction. Special contractor's tape (*Figure 2*) or sealants compatible with the wrap are used to seal the edges and joints.

WARNING!

Some building wraps are slippery and should not be used in any application where people might walk on them. Because the surface is slippery, use pump jacks or scaffolding for exterior work above the lower floor. If ladders are used, extra precautions must be taken to prevent them from sliding on the wrap.

Figure 2 ◆ Building wrap accessories.

While you must always refer to the manufacturer's instructions for specific installation information, house or building wrap is generally installed as follows:

Step 1 Using two people and beginning at a bottom corner on one side of the building, leave from 6 to 12 inches of the wrap extended beyond the corner to be used as an overlap on the adjacent side of the structure (*Figure 3*). Align the roll vertically, and unroll it for a short distance. Check that the stud marks on the wrap align with the studs of the structure. Also, check that the bottom edge of the wrap extends over and runs along the line of the foundation. Secure the wrap to the corner at 12- to 18-inch intervals.

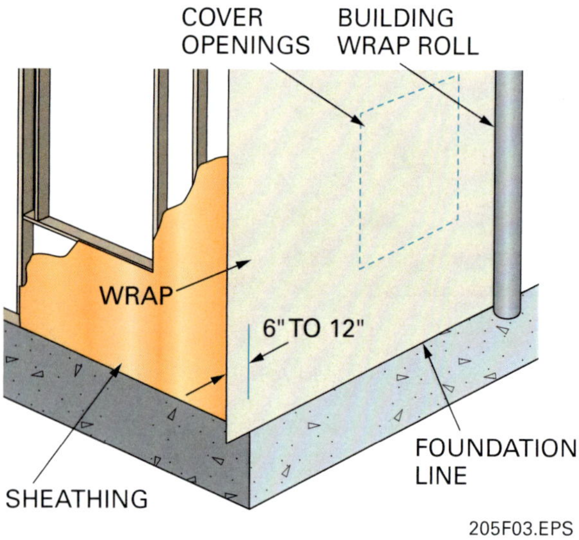

Figure 3 ◆ Starting a roll of building wrap.

Step 2 Unroll the wrap two or three more feet and make sure that it overlaps and runs along the line of the foundation. Secure the wrap vertically at 12- to 18-inch intervals on each stud using the stud marks as a fastening guide. Continue around the structure, covering all openings. If a new roll is started, overlap the end of the previous roll 6 to 12 inches to align the stud marks of the new roll with the studs of the structure.

CAUTION

Avoid any penetrations of the covering. Penetrations for door and windows must be properly sealed to avoid the entrance of water, as described later in this procedure.

Step 3 If the upper parts of the structure require coverage, repeat Steps 1 and 2, starting above the existing wrap. Make sure that the bottom edge of this layer of wrap aligns along the top edge of the lower wrap and overlaps it by 6 to 12 inches. Sometimes a building requires a double layer of wrap. When this happens, a half-width of wrap is used for the layer closest to the foundation. The first full width of wrap is installed as described in Steps 1 and 2 so that it entirely overlaps the first layer. Then, the next layer is installed as in Step 3, except that it overlaps the second layer as shown in *Figure 4*.

Step 4 At the top plate, make sure that the wrap covers both the lower and upper (double) top plate (*Figure 5*), but leave the flap loose for the time being.

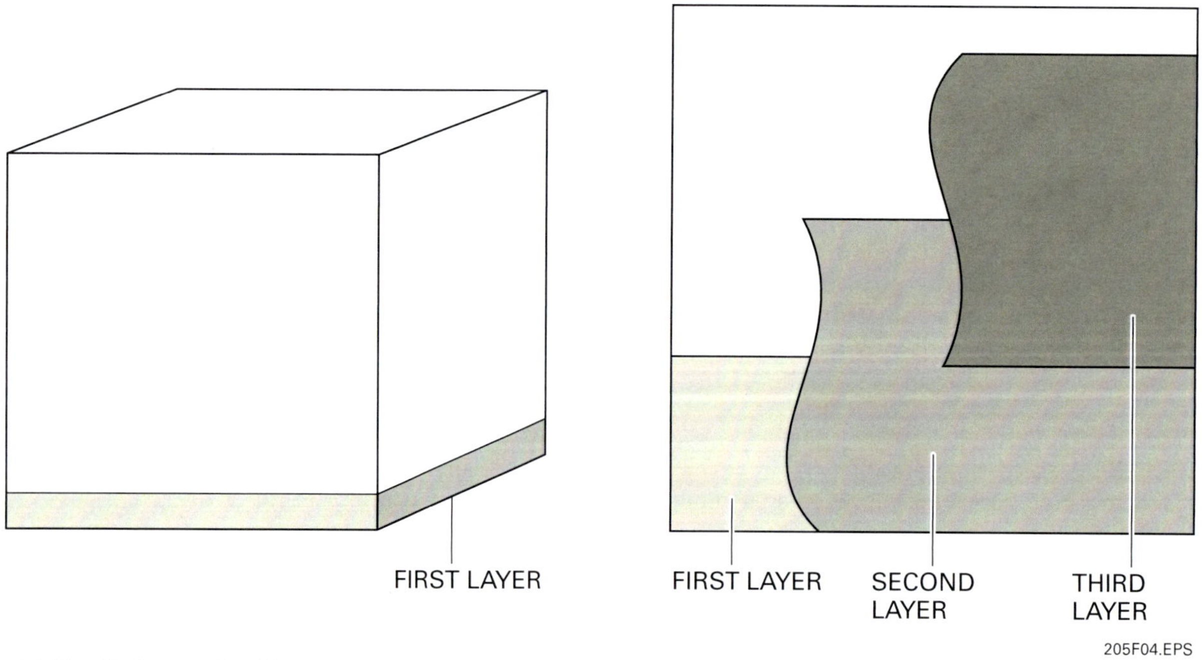

Figure 4 ◆ Double layer of building wrap.

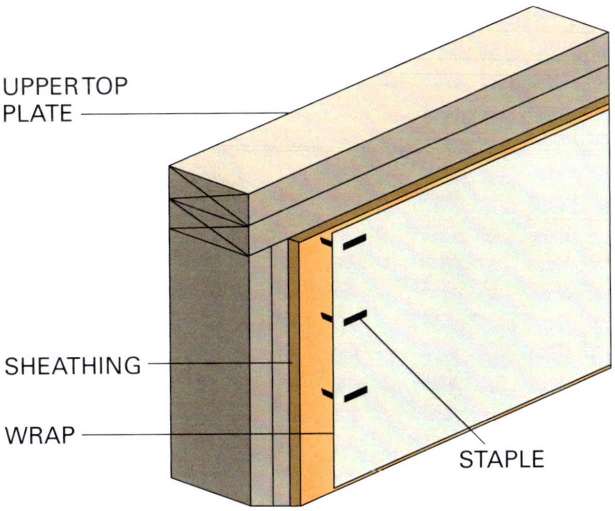

Figure 5 ◆ Top plate detail.

Step 5 At each opening, use one of the following two methods to cut back the wrap.

NOTE

Always follow the window or door manufacturer's recommendations for flashing windows or doors.

Method 1 – Uninstalled Windows/Doors:

- At the opening, cut the wrap as shown in *Figure 6*. Fold the three flaps around the sides and bottom of the opening, and secure every 6 inches. Trim off the excess.
- At the outside, install 6-inch flashing along the bottom of the opening, then install it up the sides over the top of the wrap.
- Install head flashing at the top of the opening under the wrap and over the side flashing (*Figure 7*). Tape the ends of the flaps to the head flashing using tape approved by the manufacturer.

CAUTION

It is critical that all flashing be designed and installed to shed moisture. Failure to do so will result in moisture buildup, possible mold growth, and building damage.

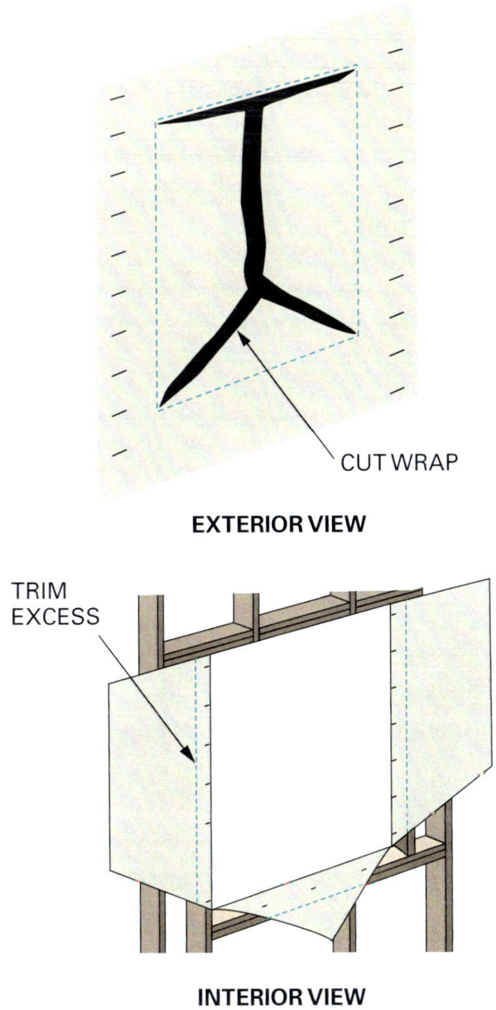

Figure 6 ◆ Cutting and folding wrap at an opening.

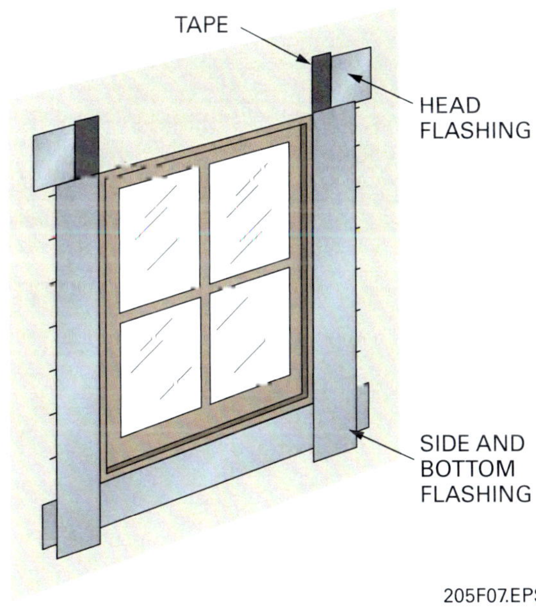

Figure 7 ◆ Installing flashing around an opening.

Method 2 – Windows/Doors with Flanges:

- Create a top flap of the wrap. Insert a head flashing under the flap and over the flange.
- Extend the flashing to the sides about 4 inches, and tape the flap to the head flashing.
- On the remaining sides, trim the wrap to overlap the flange area, and tape the edge of the wrap to the flanges (*Figure 8*).

Step 6 Secure the bottom edges of the wrap to the foundation with the recommended joint sealer, and then fasten the lower edge to the sill. At the top plate, use the sealer to seal the edge to the upper plate, and then fasten the edge to the plate.

Step 7 Seal all vertical and horizontal joints in the wrap with the recommended tape.

Step 8 Before applying the siding, repair any damage or tears in the wrap with tape or sealant.

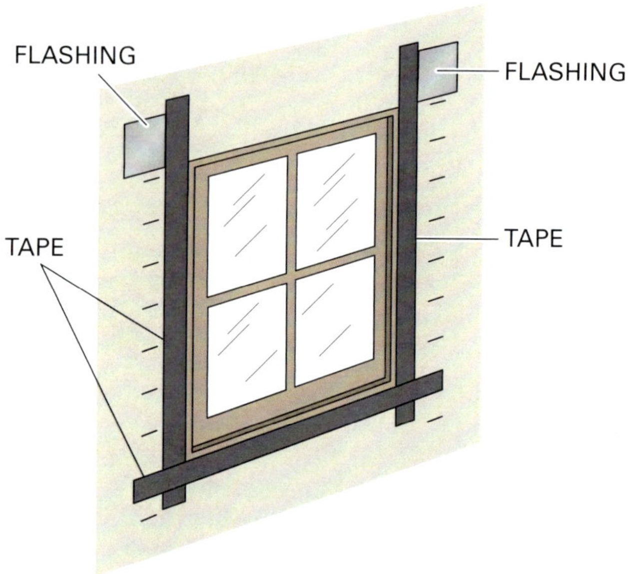

Figure 8 ◆ Installing wrap with a window in place.

INSIDE TRACK

Window Flashing System

Take special care to prevent moisture from entering the structure around windows doors. The Dupont® flashing system is designed for this purpose.

3.0.0 ♦ TRIM

A variety of trim pieces are needed to properly install exterior cladding. Manufacturers frequently make proprietary trim to accompany their cladding systems. When this is the case, it is very important to use the manufacturer's recommended trim. The basic trim pieces and their functions are as follows:

- **Control joints** – Used with stucco and EIFS to relieve stress in the coating due to thermal expansion and contraction. Control joints also help to control cracks caused by these thermal changes (*Figure 9*). With conventional stucco, the control joints must be designed to coordinate with the drainage system, with particular attention paid to splices and intersections.
- **Weep screeds** – Used with stucco and other cladding systems to drain water from behind the cladding (*Figure 10*).
- **Starter strips** – Used with EIFS to support the insulated panel and to allow water to drain from behind the cladding. Other cladding systems also use starter strips, but the design is different. When installing a starter strip, use a level and a snap line. Snap a leveling line at the installation level, and then install the starter strip on the chalkline (*Figure 11*).
- *Flashing* – Used with all cladding systems. Flashings are thin pieces of metal or flexible rubberized material that are installed above openings in the cladding system to prevent water from seeping behind the cladding.

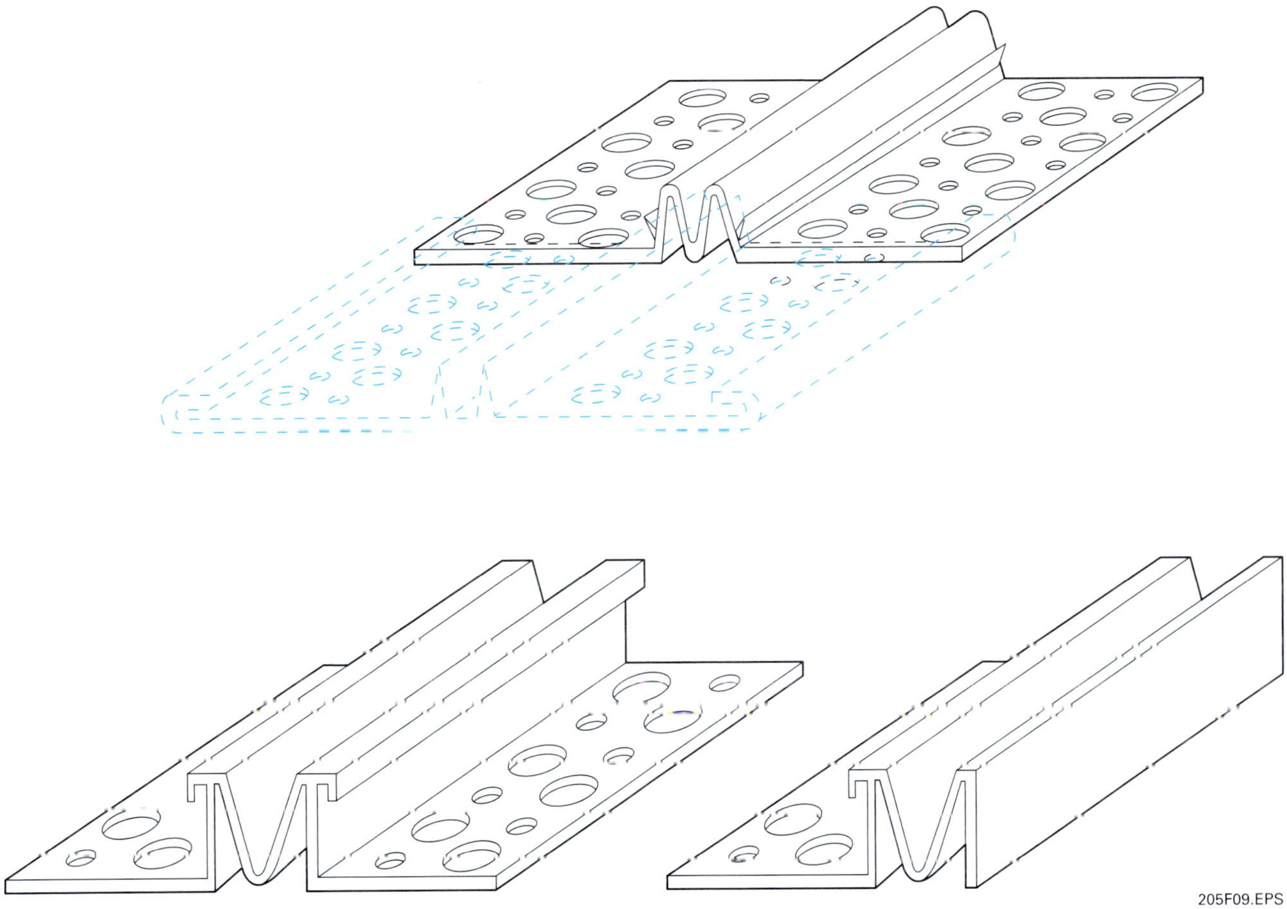

Figure 9 ♦ Control joints.

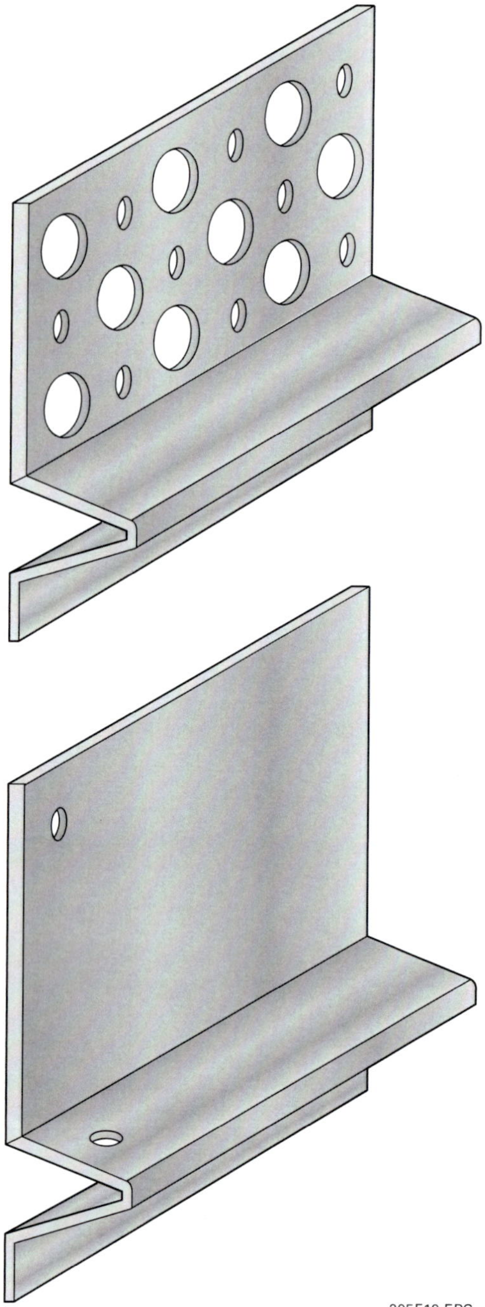

Figure 10 ◆ Weep screeds.

Flashings are installed at a slope that causes water to flow away from the opening. Improperly installed or omitted flashing is one of the causes of cladding failure (*Figure 12*).

4.0.0 ◆ CLADDING

In addition to improving a building's appearance, cladding on a structure is the primary barrier to the elements. The main focus of the different cladding systems commonly used today is to encap-

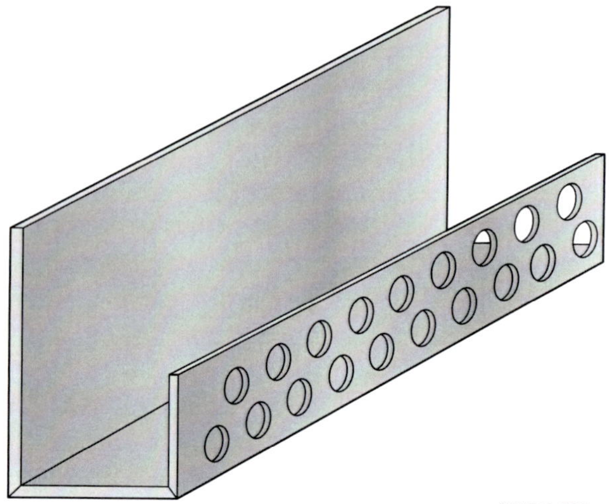

Figure 11 ◆ Starter strip.

sulate buildings with an airtight and watertight barrier. This section covers proper installation of the following cladding systems:

- Stucco
- Exterior insulation and finish systems (EIFS)
- Synthetic veneer stone
- Panelized cladding
- Glass fiber reinforced concrete (GFRC)

4.1.0 Stucco

Traditional stucco is a durable, cement-based coating for exterior walls that is breathable and allows moisture to pass through. Acrylic stucco is a synthetic material that provides a waterproof finish. Stucco can be applied over wood sheathing, fiberglass mat gypsum, or cement board, as well as over cement block and poured concrete walls. When applying stucco, building paper, such as Tyvek®, Commercial Wrap®, or Home Wrap® must be placed over the sheathing of wood frame structures. However, it is not needed over cement block or poured concrete walls.

The wrap is covered with 27" × 96" pieces of metal reinforcement mesh, called diamond mesh or lath, as recommended by the manufacturer of the stucco. The mesh is applied using special spacing fasteners or on top of a plastic mesh material that holds it slightly away from the wall. This ensures a space for water drainage.

Traditional and acrylic stucco finishes tend to expand and contract with temperature variations. These changes can cause cracks in the surface, so expansion and control joints are used. It is important to understand that these joints do not prevent cracks; they merely control their location and

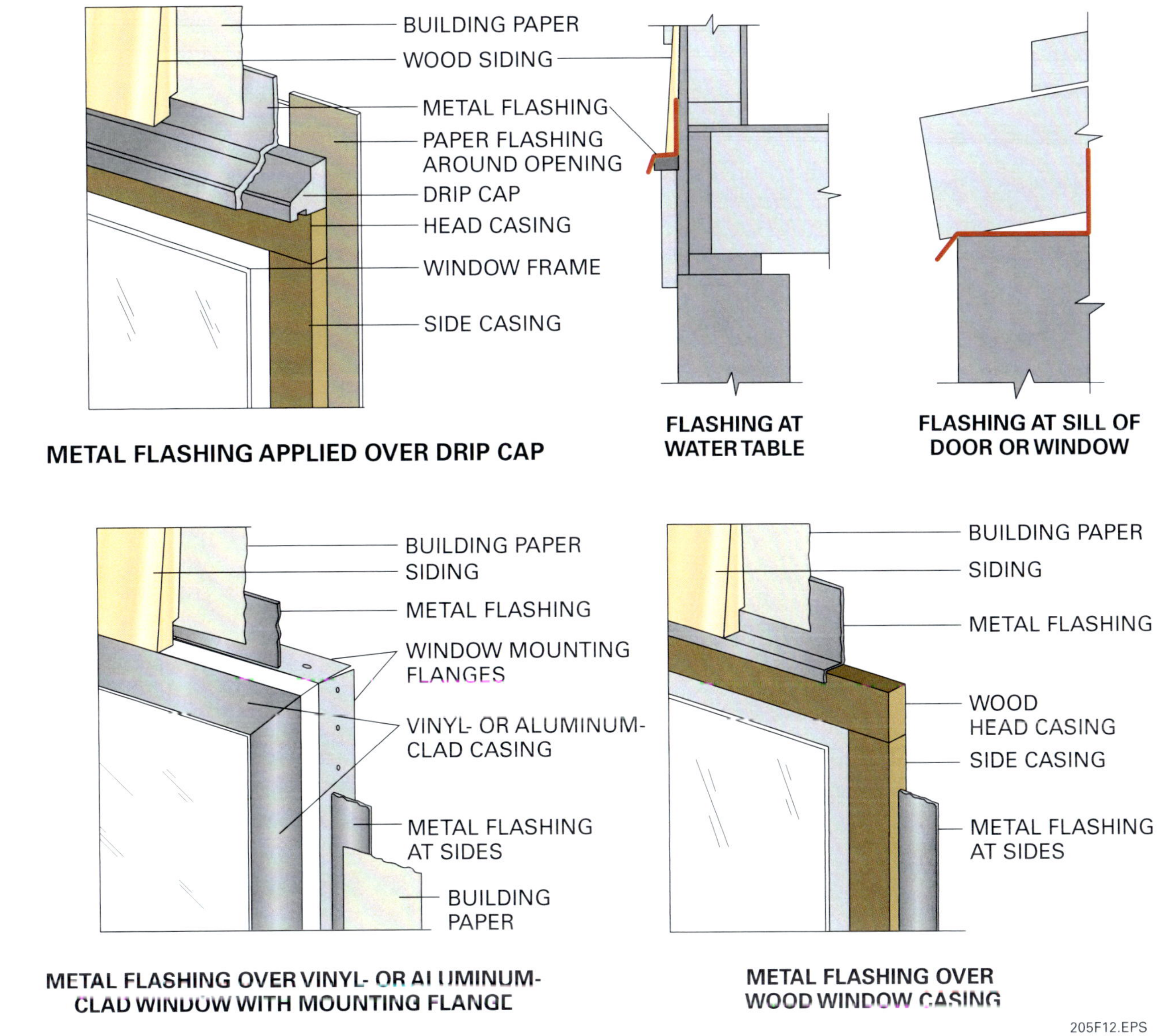

Figure 12 ◆ Flashing.

shape. The construction documents will specify the placement of expansion and control joints. Typically, they are placed at corners, around openings such as doors and windows, and at mesh joints. Placing expansion and contraction joints near doors and windows helps to disguise cracks. This is because the door or window breaks the expanse of stucco, and the joints can be hidden by trim.

> **NOTE**
> All control joints must be properly flashed as a part of the complete drainage system.

For a stucco finish, three coats of cement plaster must be applied (*Figure 13*). The first coat, called the scratch coat, is applied about ¼ to ½ inch thick over the mesh. After the scratch coat sets, but has not hardened, it is scratched with a scratch rake. The scratches should be about ⅛ inch deep. The scratches help the next layer adhere to the scratch coat. The scratch coat must be allowed to harden according to the manufacturer's specifications (usually 36 to 48 hours).

The next coat is called the brown coat. Before applying this coat, the scratch coat must be sprayed with clean water. This helps the brown coat adhere to the scratch coat. The first two coats can be applied with a rough finish using a trowel.

INSIDE TRACK

Stucco Drying Time

Stucco must be allowed to dry between coats. The amount of drying time required depends on the environmental conditions at the time the coats are applied. Refer to the manufacturer's instructions for the recommended drying/curing times.

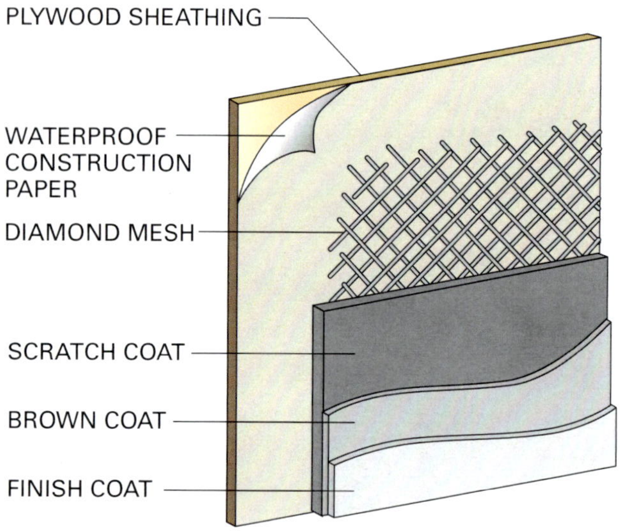

Figure 13 ◆ Stucco section.

The final coat, or finish coat, can be applied rough or smooth as desired using a variety of finishing techniques. The total thickness of the three coats should be about 1 inch, according to the Portland Cement Association.

The scratch, brown, and finish coats are a mixture of cement, sand, and clean water in different proportions. The *Standard Specification for Application of Portland Cement-Based Plaster Proportions (ASTM C 926)* states the following proportions for the coats:

- *Scratch coat* – 1 part cement to 2¼ to 4 parts sand
- *Brown coat* – 1 part cement to 3 to 5 parts sand
- *Finish coat* – 1 part cement to 1½ to 3 parts sand

Wood trim, such as frieze board or half timbers, is sometimes applied next to stucco to simulate a particular style of architecture. In such applications, the wood trim should have a rabbet—the back joining edge, called a stucco lock (*Figure 14*). When properly designed and installed, the stucco lock prevents water penetration around the stucco at the point where it joins the wood trim. When applying stucco to concrete block walls, a water barrier and mesh are not necessary, but all other requirements apply.

Stucco needs to be protected from freezing for the first 48 hours after application because the water in it can freeze and cause cracks in the finish. Stucco should be cured according to manufacturer's instructions, usually at temperatures above 40°F. In addition, recently-applied stucco may dry out too quickly in conditions of high temperature, low humidity, and wind.

4.2.0 Synthetic Stone

Synthetic stone is often used in place of real stone (*Figure 15*). Synthetic stone looks like and is installed like real stone. However, working with synthetic stone avoids many of the difficulties of real stone. Real stones are heavy; are different in color, size, and shape; and can be smooth or jagged. Typically, real stones must be trimmed and smoothed before they can be used as cladding.

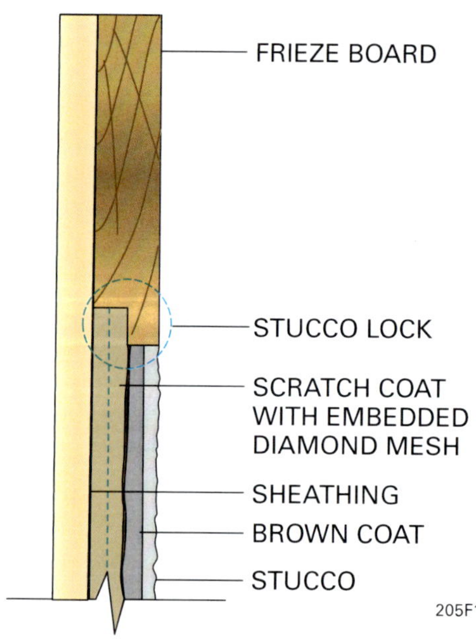

Figure 14 ◆ Stucco lock.

Figure 15 ◆ Synthetic stone.

Synthetic stone comes in many different styles, such as fieldstone and ledgestone. This variety allows designers to achieve almost any desired look they want—rustic, contemporary, traditional, or elegant. Synthetic stone is an attractive finish for frame or concrete structures, and it can be used on exteriors and interiors.

Synthetic stones are man-made and are manufactured to exact specifications. While the stones appear random in size, shape, and color, they are actually engineered to look like real stones, but are easier to install than real stones.

Wall preparation for the installation of synthetic stone is typically the same as for stucco. Some types of synthetic stones are fastened to the wall with hardware or adhesive. Other types are placed with grout, similar to real stones. Once all stones have been set, the spaces between them are filled with grout. As with all cladding, it is very important to seal any gaps around openings, such as doors and windows, to prevent water leaks.

Installing stone work is not as fast as other cladding methods. It takes a great deal of careful preparation. Before starting any synthetic stone cladding job, perform the following procedures:

- Carefully read the manufacturer's instructions.
- Lay out the stones on the ground in a clean area near the installation site. If you are working in an area where the stones can get dirty, use scrap plywood or other material to lay out the stones.
- Mix stones from different boxes so that the variations are randomly dispersed.
- Lay out the stones with no more than a ½-inch gap between them.
- Lay out the stones in an attractive pattern mixing sizes, colors, and shapes.
- Avoid placing stones in ways that will require long grout lines. Use different-sized stones to break up patterns.
- At inside corners, alternate placing stones on each wall, so they can be woven together. Use an alternating long-short pattern, cutting the stones as needed to ensure a proper fit. Be sure to cut the stone so that the cut side is facing the corner.
- Some manufacturers make trim pieces that can be used on outside corners and around doors and windows. When using these pieces, place them first, and then place the stones to the trim. Use small stones to avoid having to cut pieces to fit.
- When you need to cut a stone to fit it, check the manufacturer's recommendations about how to cut the stone and what personal protective equipment you should use. Be sure to clean all dust from the stone before you set it. Any dust on the stone can prevent a good bond with the grout.
- When you need to cut a stone to fit it, always cut it so the cut side is not visible. That is, when the stone is above eye level, set the cut side facing up; when the stone is below eye level, set the cut side facing down.
- Do not cut stones that will be set at eye level.
- Determine if stones and grout are to be sealed. Sometimes sealing is required if the stone will be in an area where soil can be splashed onto its surface, such as near a flower bed. Always keep in mind that sealing the stones and grout can darken the color.

When you are ready to place the stones, secure them according to the manufacturer's specification. The following procedure describes setting stones with grout:

Step 1 Mix the grout according to the manufacturer's instructions, adding coloring as desired. Always mix the grout and coloring agent in the same proportions to avoid different shades of color. Mix only enough grout that you can use before it sets. Never use partially set grout. This can cause the installation to fail.

Step 2 When instructed to do so by the manufacturer, moisten the brown coat with clean water. This helps the grout to adhere to the base coat. This process may need to be repeated periodically when the weather is warm and dry or when the installation area is in direct sunlight.

Step 3 Determine the direction in which the stones will be installed. Some products need to be installed from the top down. This helps to keep the stones clean of spattered grout. Some products should be installed from the bottom up. This helps to keep the stone line level.

Step 4 Once you have established a starting point, use a level and a snap-line and snap a leveling line at the starting level. When installing flat stones, such as brick and ledgestone, snap chalklines every 8 to 12 inches above the starting level. This will help to keep a level installation.

Step 5 Dust off the stones using a clean whisk broom. Always clean all dirt, sand, and dust from the stones before placing them. This will ensure a good bond with the grout.

Step 6 Using a metal trowel, spread about ½ inch of grout on the back of the stone (*Figure 16*). Set the stone roughly ½ inch from its neighboring stones. Firmly press the stone into place until grout oozes from the back of it. Firm placement is needed to get a good bond. An alternate method is to spread about ½ inch of grout over the brown coat, and then firmly place the stones until the area is covered. When using this method, spread the grout over a small area of roughly 4 feet by 4 feet. This technique helps prevent the grout from drying out before the stones are set.

205F16.EPS

Figure 16 ◆ Applying grout to a synthetic stone.

NOTE

Never try to wipe away wet grout from synthetic stone. This can smear the grout and permanently stain the stone. Instead, wait until the grout is dry, and then brush it away with a clean whisk broom.

Some manufacturers recommend dampening the back of the stone to help the grout adhere to it. When this is the case, be sure that the stone is not so wet that water puddles on its surface. This will prevent the grout from adhering to the stone.

Step 7 Once all stone in the area has been set, use a caulk gun or fill a grout bag about two-thirds full of grout, and twist the top to keep it closed. Carefully squeeze the grout into the cracks between the stones. Take care that the joint line is free of gaps and the grout is free of air bubbles.

Step 8 Allow the grout to firm but not set, and then finish the grout with a striking tool to remove any excess, leaving a concave surface.

Step 9 Seal the grout and stones if this is recommended by the manufacturer.

4.3.0 Exterior Insulation and Finish Systems (EIFS)

Exterior insulation and finish systems (or EIFS) combine insulation with a synthetic, textured, exterior coating. This finishing system brings together the following: an insulating panel of expanded polystyrene, which is covered with waterproof membrane; mesh; a base coat; and a watertight finish coat. The watertight finish coat, called synthetic stucco, is an elastomeric coating. This cladding looks attractive and provides good insulation from both heat and cold, which helps lower heating and cooling costs (*Figure 17*).

There are two types of EIFS systems: traditional and water managed. Traditional systems rely on the insulation and coating to prevent water infiltration. This expectation is unrealistic. Even condensation from thermal changes can cause water to collect behind the panel, causing wood surfaces to rot and encouraging mold to grow. Today, water managed systems are used.

The latest system uses water barriers to prevent leaks. Proper installation of EIFS is critical. It is extremely important to follow the manufacturer's instructions when installing EIFS to prevent voiding any warranties. Use only manufacturer-recommended barriers, adhesives, and components.

EIFS is a multilayer system similar to those used in stucco finishes (*Figure 18*). The most critical layer is the water management layer. Any walls that have exterior wood, including masonry or concrete buildings with wood furring, must have a waterproof exterior to prevent water from contacting the wood. Wood-framed buildings are covered with an appropriate moisture barrier. In addition, flashing must be installed over doors, windows, and on any other penetrations that can allow water to enter. If any water penetrates the moisture barrier, it is drained out the bottom of the system. For this process to be effective, there can be no tears, openings, or breaks in the barrier.

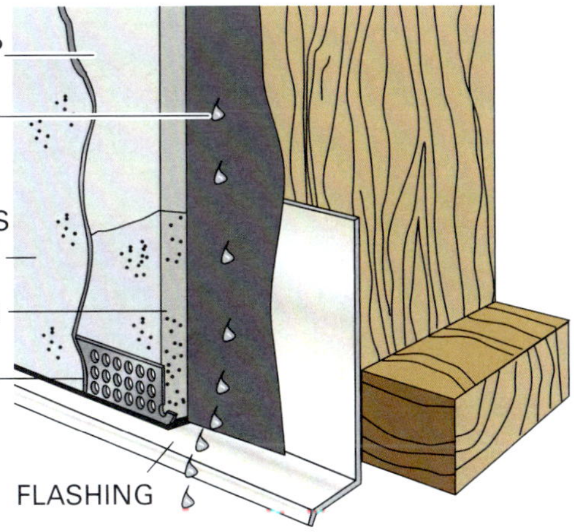

205F18.EPS

Figure 18 ◆ Typical EIFS installation.

205F17.EPS

Figure 17 ◆ EIFS.

> **NOTE**
>
> Masonry and concrete structures with no exposed wood do not require a water barrier.

A layer of insulation is applied over the water management layer. Panels of expanded polystyrene are usually used as the insulation in this system. Occasionally, cement board, such as Durock®, is used. Polystyrene is commonly used as building insulation. Expanded polystyrene (EPS) is made from small pellets of polystyrene that have been molded together into a particular shape—panels, in this case. Extruded polystyrene (XPS) is a foam of polystyrene that is pressed into a mold. XPS is sold under the brand name Styrofoam™. The term Styrofoam™ is often mistakenly used to describe EPS. When polystyrene is made up of pellets—white foam coffee cups for example—it is EPS. XPS is fine celled foam. Both EPS and XPS have low melting points and can be easily cut in the field. XPS requires the use of control joints and mechanical fasteners. It often cut into architectural trim and other shapes, such as those shown in *Figure 19*.

The insulation panels are joined with sealant to form a watertight seal. A rasp is used to provide a uniform, flat surface. The panels are backwrapped with a mesh and basecoat (*Figure 20*) to protect the edges from water. Panels adjacent to door or window jambs must be backwrapped and then caulked with the recommended sealant to prevent leaks (*Figure 21*). The base coat should be primed to ensure a good finish coat.

EIFS elastomeric finish can be applied to stucco. When finishing EIFS, it is important to keep a wet edge. When installing large areas, provide joints within the finished surface. This provides separate blocks that allow for stopping points during installation and help in maintenance, as you only need to refinish one block.

> **DID YOU KNOW?**
>
> **Exterior Insulation and Finish Systems**
>
> EIFS were originally developed as a method of repairing historical buildings in post-war Europe.

Figure 19 ◆ Architectural trim.

Once installed, it is often difficult for even an experienced inspector to distinguish stucco from EIFS. One telltale sign is that EIFS sounds hollow when tapped, but stucco sounds solid.

When properly installed, EIFS can provide years of good appearance and service. EIFS relies on its ability to maintain a watertight seal over the wooden framing. Any failure in the wrapping can result in serious water damage. Most EIFS failures are due to improper installation of the wrap, the flashings, or the system itself. One type of EIFS failure that is unrelated to installation occurs when the wooden framing dries out and the building moves, creating openings in the system.

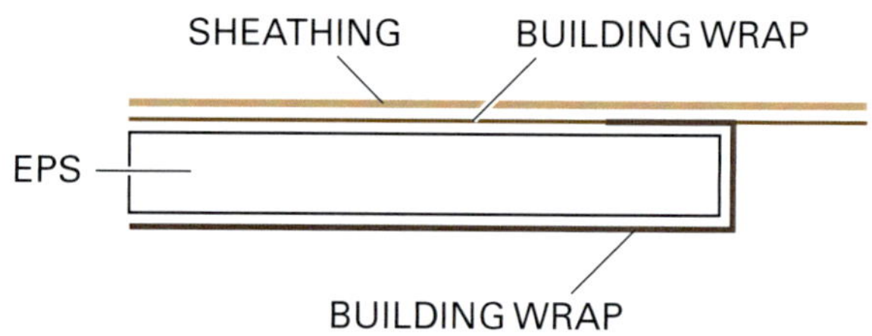

Figure 20 ◆ Backwrapping.

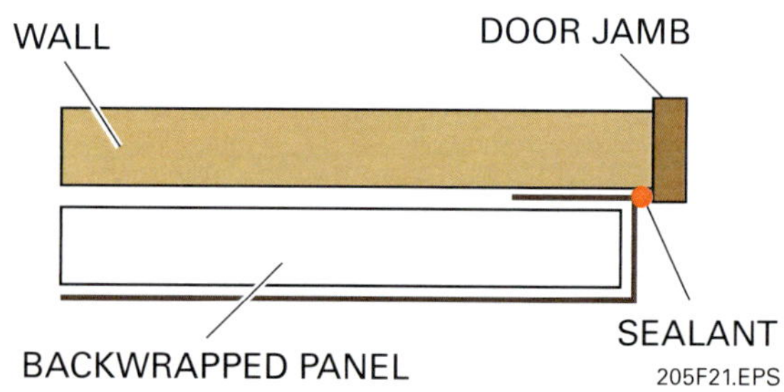

Figure 21 ◆ Door jamb.

4.4.0 Panelized Cladding

Fiber-cement siding is a man-made product that is made using portland cement, sand, fiberglass and/or cellulose fiber, selected additives, and water. It is usually pressure formed and heat cured. This material can be molded into a variety of shapes, including panels.

The major advantages fiber-cement siding has over hardboard or wood siding are that it will not rot, is noncombustible, and can withstand a termite attack. Fiber-cement siding is highly resistant to impact damage and, in some cases, can withstand hurricane-force winds of 130 mph or more. It also resists permanent damage from water and salt spray. This siding is especially well suited for use in areas prone to fires or high winds.

As with wood siding, fiber-cement siding is available in single-lap siding in widths from 6 to

INSIDE TRACK

Fiber-Cement Siding Styles

A number of architectural styles of fiber-cement siding are available and can be used to obtain different effects. Planks are available as smooth or wood grained, and panels are available as plain, stucco, and vertical wood grained.

HARDIPANEL® SMOOTH VERTICAL SIDING

HARDIPANEL® STUCCO VERTICAL SIDING

HARDIPANEL® SIERRA-8 VERTICAL SIDING

HARDIPANEL® CEDARMILL VERTICAL SIDING

205SA02.EPS

12 inches, and as vertical panels. Commercial buildings are more likely to use the panels as cladding because they can be installed quickly and have fewer joints. The panels come in a number of different surface patterns. The recommended finish is 100 percent acrylic latex paint over an alkali-resistant primer. However, gloss or satin oil/alkyd paints over an alkali-resistant primer may also be used. The panels cannot be stained.

To achieve satisfactory performance, and for warranty purposes, the siding must be installed and finished as specified by the manufacturer. The following are general guidelines for installing fiber-cement siding:

- The siding may be applied over walls sheathed with wood or insulation board up to 1 inch thick and with studs spaced not more than 24 inches OC. The lowest edge of the siding should not be in contact with the earth or standing water. When cutting the siding, be sure to prime all cut ends with an alkali-resistant primer.
- Store the panels flat, and keep them dry and covered while in storage. Avoid installing wet panels. Wet panels swell, and as they dry they shrink. Due to this shrinkage, the joints of panels that are installed wet will pull apart.
- To avoid bending the panels, carry them upright and on edge. Use a two-person carry as necessary. The panels are durable, but the corners and edges are prone to breakage.
- A weather-resistant barrier, such as Tyvek® or HardieWrap™, may be required under the siding when it is applied directly to studs or over wood sheathing. Consult the manufacturer's instructions for specific requirements.
- Fiber-cement siding may be cut using the following tools: a power saw with a fine-toothed, carbide-tipped or dry-diamond circular saw blade; electric or pneumatic carbide-tipped power shears (*Figure 22*); or a score-and-snap knife with a tungsten-carbide tip.

WARNING!

Because this dry material will be drilled, cut, and/or abraded, proper respiratory protection must always be used to avoid inhaling toxic silica dust that can cause a fatal lung disease called silicosis.

- Only galvanized steel, copper, or stainless steel flashing and screws/nails may be used when installing fiber-cement siding. The size and style of fasteners used on a job depends on the height of the building, the maximum wind speed in the area, and the use of the building. Consult with the manufacturer of the siding to determine the correct fasteners for the job. Never use staples.
- Galvanized steel with a powder or baked enamel finish or vinyl inside/outside corners and other trim can be used and painted to match the siding finish. Never use aluminum trim components or fasteners; they will corrode when in contact with the siding.
- At horizontal joints, maintain a ¼-inch space between panels, and install Z-flashing.
- Maintain a minimum of 1 inch to 2 inches between the bottom of the panel and decking, steps, driveways, or other surfaces, excluding the finished grade. Maintain a minimum of 6 inches between the bottom of the panel and the finished grade.
- Always prime the product per the manufacturer's instructions.

Details of typical panel siding installation are shown in *Figure 23*.

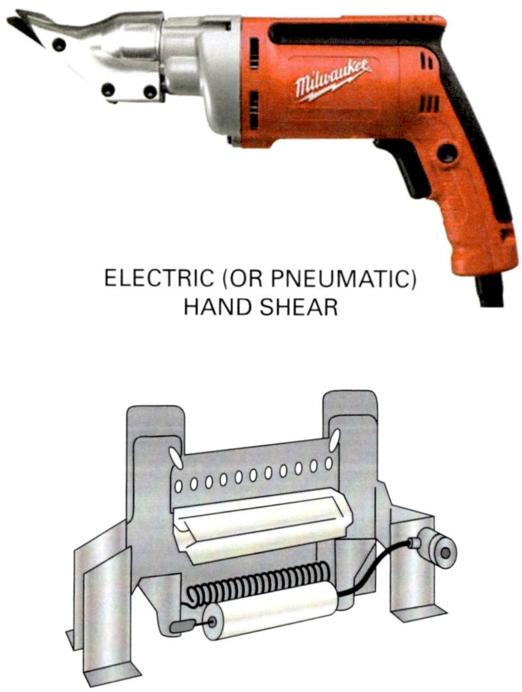

Figure 22 ◆ Power shears.

Hardipanel® *Vertical Siding*

James Hardie Building Products

INSTALLATION INSTRUCTIONS PRIMED AND COLORPLUS® TECHNOLOGY TREATED PRODUCTS — EFFECTIVE DECEMBER 2005

SMOOTH * CEDARMILL© * SELECT SIERRA 8" * STUCCO

IMPORTANT: FAILURE TO INSTALL AND FINISH THIS PRODUCT IN ACCORDANCE WITH APPLICABLE BUILDING CODES AND JAMES HARDIE'S WRITTEN APPLICATION INSTRUCTIONS MAY LEAD TO PERSONAL INJURY, AFFECT SYSTEM PERFORMANCE, VIOLATE LOCAL BUILDING CODES, AND VOID THE PRODUCT ONLY WARRANTY.

STORAGE & HANDLING:

Store flat and keep dry and covered prior to installation. Installing siding wet or saturated may result in shrinkage at butt joints. Carry panels on edge. Protect edges and corners from breakage. James Hardie is not responsible for damage caused by improper storage and handling of the product

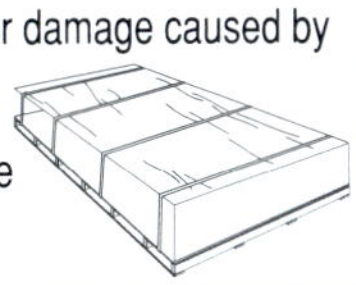

⚠ CUTTING INSTRUCTIONS

OUTDOORS	INDOORS
1. Position cutting station so that wind will blow dust away from user and others in working area. 2. Use one of the following methods: a. Best: i. Score and snap ii. Shears (Pneumatic or Handheld) b. Better: i. Dust reducing circular saw equipped with a Hardiblade® saw blade and HEPA vacuum extraction c. Good: i. Dust reducing circular saw with a Hardiblade saw blade (only use for low to moderate cutting)	1. Cut only using score and snap, or shears (manual, electric or pneumatic). 2. Position cutting station in well-ventilated area - NEVER use a power saw indoors - NEVER use a circular saw blade that does not carry the Hardiblade saw blade trademark - NEVER dry sweep – Use wet suppression or HEPA Vacuum

Important Note: For maximum protection (lowest respirable dust production), James Hardie recommends always using "Best"-level cutting methods where feasible.

NIOSH-approved respirators can be used in conjunction with above cutting practices to further reduce dust exposures. Additional exposure information is available at www.jameshardie.com to help you determine the most appropriate cutting method for your job requirements. If concern still exists about exposure levels or you do not comply with the above practices, you should always consult a qualified industrial hygienist or contact James Hardie for further information.

SD083105

GENERAL REQUIREMENTS:

- Hardipanel® siding can be installed to studs at maximum 24" o.c. Irregularities in framing and sheathing can mirror through the finished application. Refer to relevant compliance documents and Fastener Requirements' for specific arrangement and type of fasteners for your application.
- A Weather-resistive barrier is required in accordance with local building code requirements. The weather resistive barrier must be appropriately installed with penetration and junction flashings in accordance with local building code requirements. James Hardie will assume no responsibility for water infiltration.
- Install James Hardie® products with a minimum 6" clearance to the finished grade on the exterior of the building or in accordance with local building codes if greater than 6" is required (fig. 4).
- Maintain a minimum 2" clearance between James Hardie products and roofs, decks, paths, steps and driveways (fig. 5 & 6).
- Maintain a 1/4" clearance between James Hardie products and horizontal flashing (fig. 7).
- Install kickout flashing at roof-wall junctions. (fig 8.) Ensure gutters have end caps and do not terminate against siding and trim.
- Adjacent finished grade must slope away from the building in accordance with local building codes - typically a minimum of 6" in the first 10'.
- Do not install James Hardie products, such that they may remain in contact with standing water.

INSTALLATION:

Fastener Requirements

Position fasteners 3/8" from panel edges and and no closer than 2" away from corners. Do not nail into corners.

Hardipanel® Siding Installation

- Framing must be provided at horizontal and vertical edges for nailing.
- Hardipanel® siding must be joined on stud.

Joint Treatment

- Vertical Joints - Install panels in moderate contact (fig. 1), Alternatively joints may also be covered with battens, PVC or metal jointers or caulked (fig. 2).
- Horizontal Joints - Provide Z-flashing at all horizontal joints

Figure 1

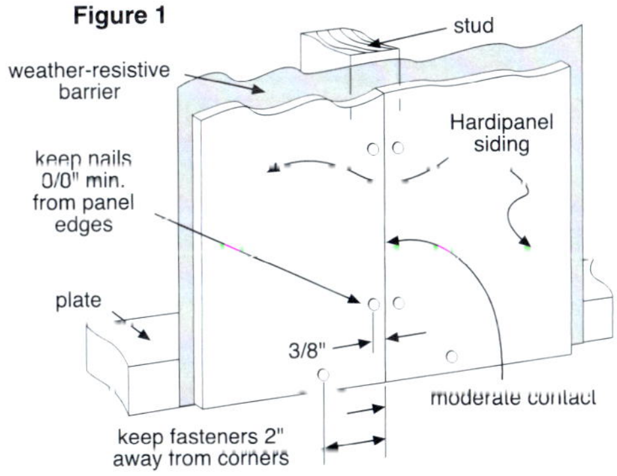

Figure 2

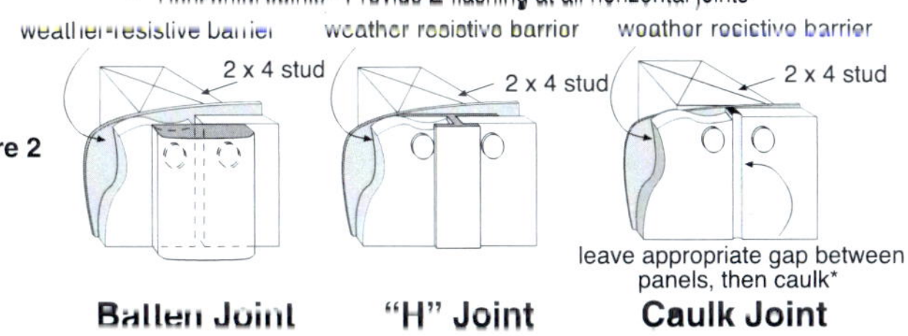

Figure 3

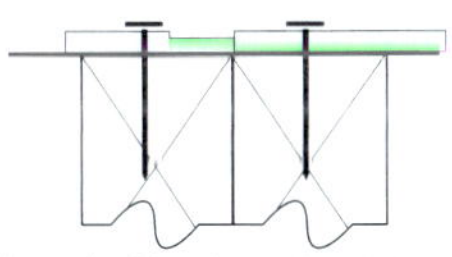

Recommendation: When using Sierra 8, provide a double stud at panel joints to avoid nailing through grooves.

*Apply caulk in accordance with caulk manufacturers written application instructions. Double stud may be required to maintain minimum edge nailing distances.

WARNING: AVOID BREATHING SILICA DUST

James Hardie® products contain respirable crystalline silica, which is known to the State of California to cause cancer and is considered by IARC and NIOSH to be a cause of cancer from some occupational sources. Breathing excessive amounts of respirable silica dust can also cause a disabling and potentially fatal lung disease called silicosis, and has been linked with other diseases. Some studies suggest smoking may increase these risks. During installation or handling: (1) work in outdoor areas with ample ventilation; (2) use fiber cement shears for cutting or, where not feasible, use a Hardiblade® saw blade and dust-reducing circular saw attached to a HEPA vacuum; (3) warn others in the immediate area; (4) wear a properly-fitted, NIOSH-approved dust mask or respirator (e.g. N-95) in accordance with applicable government regulations and manufacturer instructions to further limit respirable silica exposures. During clean-up, use HEPA vacuums or wet cleanup methods - never dry sweep. For further information, refer to our installation instructions and Material Safety Data Sheet available at www.jameshardie.com or by calling 1-800-9HARDIE (1-800-942-7343). FAILURE TO ADHERE TO OUR WARNINGS, MSDS, AND INSTALLATION INSTRUCTIONS MAY LEAD TO SERIOUS PERSONAL INJURY OR DEATH.

205F23.EPS

Figure 23 ◆ Typical fiber-cement panel siding installation details.

4.5.0 Glass Fiber Reinforced Concrete (GFRC)

Glass fiber reinforced concrete (GFRC) is fairly new as an exterior cladding. It is a concrete product that is reinforced with glass fibers that make it durable and light in weight. GFRC can be made into panels or molded into various architectural shapes (*Figure 24*). GFRC can even be produced to match sections of older buildings that have been damaged.

GFRC products can be installed over wood frame or concrete/masonry construction. When installed over wood, a moisture barrier must be applied to the structure. GFRC panels can be installed with or without insulation. Although the panels are made of concrete, they are not strong enough to be loadbearing. Since GFRC products are fabricated, they are consistent in size and color. They are available in a number of finishes, including aggregate and brick.

GFRC products may be installed with fasteners through the front of the panel or with hanger devices on the rear of the panel. All fasteners must be corrosion resistant and should match the panel color if they penetrate the front side. Panels should be installed per the manufacturer's instructions. Some producers specify the need for control joints in certain installations; others may specify joint gaps between panels. All gaps should be filled with the recommended type of sealant.

205F24.EPS

Figure 24 ◆ GFRC architectural details.

Review Questions

1. The primary purpose of exterior cladding is to _____.
   a. hide blemishes
   b. protect the structure
   c. look good
   d. provide insulation

2. When wrapping a building, the beginning of a new roll should overlap the end of the previous roll by _____ inches.
   a. 1 to 6
   b. 6 to 12
   c. 12 to 18
   d. 18 to 24

3. A half width roll of building wrap can be used _____.
   a. when a building needs to be double wrapped
   b. to wrap a building that is very short in height
   c. when one person is wrapping the building
   d. to wrap walls below the roofline

Match the trim term with the correct description of its use.

4. _____ Control joint
5. _____ Weep screed
6. _____ Starter strip
7. _____ Flashing

   a. Drains water from behind the cladding
   b. Covers vertical seams to prevent water infiltration
   c. Supports the insulated panels of EIFS
   d. Relieves coating stress due to thermal changes
   e. Prevents water from seeping behind the cladding

8. Traditional stucco is _____.
   a. waterproof
   b. cement-based
   c. obsolete
   d. man-made

9. Stucco cladding over a concrete structure is applied over _____.
   a. building wrap
   b. metal mesh
   c. scratch and brown coats
   d. chicken wire

10. Stucco finishes tend to expand and contract with temperature changes.
    a. True
    b. False

11. Expansion joints in stucco are placed _____.
    a. every 100 feet
    b. at mesh joints
    c. below grade
    d. wherever it is convenient

12. A scratch coat applied over mesh should be approximately _____ inch thick.
    a. ⅛ to ¼
    b. ¼ to ½
    c. ½ to ¾
    d. ¾ to 1

13. A brown coat is a mixture of _____.
    a. browning and cement
    b. browning and sand
    c. sand and cement
    d. sand, browning, and cement

14. When cutting synthetic stone, be sure to _____.
    a. wet the stone before you cut it
    b. grind off any rough edges
    c. clean the stone after it is cut
    d. grout the stone before you cut it

15. When placing cut stones up to an inside corner, place them _____.
    a. so the grout lines are straight
    b. in a long-short pattern
    c. cut side facing out
    d. ½ inch from the wall

Review Questions

16. To keep flat stones level when installing them, snap a chalkline every _____ inches as a guide.
    a. 4 to 8
    b. 6 to 10
    c. 8 to 12
    d. 10 to 14

17. If you spatter grout on a synthetic stone after it is placed, you should _____.
    a. wipe the grout away with a damp cloth
    b. wait for the grout to dry, and then brush it away
    c. remove the stone from the wall, and soak it in water
    d. realize that grout spatters cannot be removed from synthetic stones

18. When setting synthetic stone, use a striking tool to _____.
    a. fit the stones
    b. firmly set the stones
    c. finish the grout and remove any excess
    d. scratch the brown coat

19. Extruded polystyrene is _____.
    a. made of pellets of polystyrene molded into a shape
    b. a fine celled foam
    c. mistakenly called Styrofoam™
    d. used in coffee cups

20. One reason for EIFS failure in wooden framed buildings is that _____.
    a. the wood dries, causing the building to move
    b. the buildings are susceptible to termite damage
    c. wood swells in humid weather
    d. polystyrene breaks down quickly

21. Fiber-cement siding should be finished with _____.
    a. latex paint
    b. varnish
    c. stain
    d. cement sealer

22. An acceptable way to fasten fiber-cement siding to a stud is to use _____.
    a. staples
    b. plastic fasteners
    c. galvanized steel screws
    d. aluminum nails

23. At horizontal joints, install fiber-cement panels _____ inch apart.
    a. ⅛
    b. ¼
    c. ⅜
    d. ½

24. Fiber-cement siding should be installed _____ inch(es) above decking.
    a. ½
    b. ½ to 1
    c. 1 to 2
    d. 2

25. GFRC can be installed over insulation.
    a. True
    b. False

Summary

Exterior cladding does more than make a building look good. Its primary purpose is to protect the structural members of a building from air and water infiltration. For exterior cladding to do its job, it must be installed properly. Omitting or improperly installing small pieces of trim or forgetting to apply the correct sealant to openings can cause the cladding to fail. To do your job properly, you must carefully read, understand, and follow the cladding manufacturer's instructions.

Notes

Trade Terms Introduced in This Module

Control joint: A trim piece used with stucco and exterior insulation and finish systems to relieve stress in the coating due to thermal expansion and contraction.

Elastomeric: Having the elastic properties of rubber.

Flashing: Thin strips of material that are placed above openings in a structure to divert water away from the opening.

Permeable: Having pores or openings that permit liquids, such as water, or gases to pass through.

Starter strip: A trim piece used with exterior insulation and finish systems to support the insulated panels.

Weep screed: A trim piece used with stucco and other cladding systems to drain water from behind the cladding.

Additional Resources and References

Additional Resources

This module is intended to be a thorough resource for task training. The following reference works are suggested for further study. These are optional materials for continued education rather than for task training.

http://www.cement.org

http://www.culturedstone.com (1-800-255-1727)

http://www.eima.com

http://www.eldoradostone.com

http://www.jameshardiecommercial.com

Figure Credits

Johns Manville, 205F01 (top)

Dupont™ Building Innovations™, 205F01 (middle, bottom), 205F02, 205SA01

Owens Corning, 205F15, 205F16

Sto Corporation, 205F17

Topaz Publications, Inc., 205F19

James Hardie Building Products, 205SA02, 205F23

Milwaukee Electric Tool, 205F22 (top)

The Stromberg Group, 205F24

NCCER CURRICULA — USER UPDATE

NCCER makes every effort to keep its textbooks up-to-date and free of technical errors. We appreciate your help in this process. If you find an error, a typographical mistake, or an inaccuracy in NCCER's curricula, please fill out this form (or a photocopy), or complete the on-line form at **www.nccer.org/olf**. Be sure to include the exact module ID number, page number, a detailed description, and your recommended correction. Your input will be brought to the attention of the Authoring Team. Thank you for your assistance.

Instructors – If you have an idea for improving this textbook, or have found that additional materials were necessary to teach this module effectively, please let us know so that we may present your suggestions to the Authoring Team.

NCCER Product Development and Revision

13614 Progress Blvd., Alachua, FL 32615

Email: curriculum@nccer.org

Online: www.nccer.org/olf

❑ Trainee Guide ❑ Lesson Plans ❑ Exam ❑ PowerPoints Other ______________________

Craft / Level: Copyright Date:

Module ID Number / Title:

Section Number(s):

Description:

Recommended Correction:

Your Name:

Address:

Email: Phone:

45206-09

Specialty Finishes

45206-09
Specialty Finishes

Topics to be presented in this module include:

Overview

Interior finish work on a commercial building gives it eye appeal, character, and a distinctive look. Even a coat of paint can perk up an area and make it look more welcoming and attractive. Some specially formulated paints can give walls and ceilings a dramatic, one-of-a-kind look with very little effort. With a few simple application techniques, paint can simulate the appearance of marble, sandstone, or Venetian plaster. In addition, texturing compounds can be used to enhance the appearance of any surface. Texturing adds depth to a smooth surface, giving it a visually interesting, customized look. This module will introduce you to basic information about paints, painting, and texturing, as well as application techniques that can give interiors a rich, warm appearance.

Objectives

When you have completed this module, you will be able to do the following:

1. Identify various material types for interior finishes and state their applications.
2. Identify and select the appropriate tools used to apply specific interior finishes.
3. Identify various finish and texture samples.

Trade Terms

Cut-in
Potlife
Viscosity

Prerequisites

Before you begin this module, it is recommended that you successfully complete *Core Curriculum; Drywall Level One*; and *Drywall Level Two*, Modules 45201-09 through 45205-09.

This course map shows all of the modules in the second level of the *Drywall* curriculum. The suggested training order begins at the bottom and proceeds up. Skill levels increase as you advance on the course map. The local Training Program Sponsor may adjust the training order.

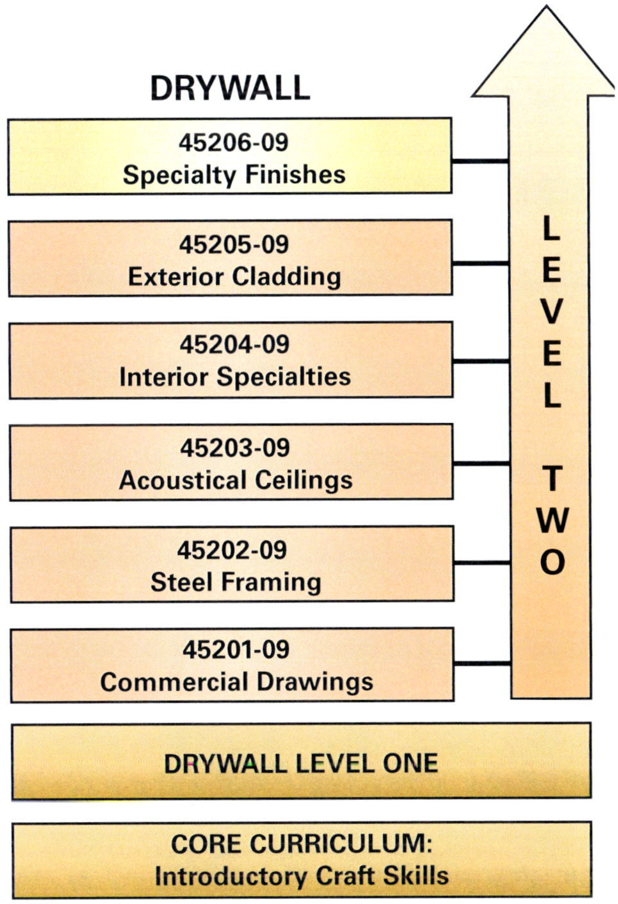

206CMAP.EPS

1.0.0 ◆ INTRODUCTION

A room's appearance may be one of the things you think of as being least important when you are working on a job. However, appearance is the first thing that owners or tenants notice. Just a simple coat of paint can enhance the look of an area and make it look attractive, clean, and professional.

In addition to painting, a number of easy paint and texture application techniques can be used to add a designer look to any drywall surface. These techniques include the following:

- *Textured surfaces* – Textured surfaces are very popular. Not only do they enhance the appearance of any surface, they also disguise surface blemishes. Many joint compounds can be used as texturing material. Other texturing materials are specially designed for this purpose. Texturing may be sprayed onto a surface or applied by hand in a desired pattern. Some texturing materials are sprayed onto the surface and then finished by hand.
- *Textured paints* – Textured paints are thicker than regular paint. When they are applied to a surface, they can be manually worked to achieve a certain look. Other paints have fine, medium, or coarse texturing material mixed into the paint. Texturing material may be in the base coat, which is usually white or off-white. It may also be mixed into the paint that will be used for the final coat. Textured paints are similar to textured surfaces in their ability to hide surface flaws. These paints may be sprayed, brushed, or rolled onto a surface in accordance with the manufacturer's instructions.
- *Faux finishes* – Faux finishes have recently gained popularity. This finishing technique uses two or more shades of paint to simulate materials such as marble, stone, leather, or suede. To achieve a faux finish, two or more layers of paint are applied using brushes, sponges, rags, or paper. For the finish to look natural, it should have a random pattern and varying shades of colors. Applying faux finishes gives you an opportunity to be creative.

2.0.0 ◆ PAINTS

Paint is an essential part of interior decorating. It is inexpensive, comes in a wide array of colors, can be applied in many different ways, and is easy to use. Paint is a mixture of several components, including the following:

- *Pigments* – Pigments give paint its color. Paint is available as a pre-mixed color, or it can be custom-colored just before purchase. Some vendors use computers to analyze a sample color, then they determine the paint formula needed to duplicate that color. Paints are colored and the desired shades achieved by mixing different colored pigments into a tint coat, which is usually white, off-white, or light beige.
- *Binders* – Also called resins, binders make the paint dry into a film. A paint that is currently very popular is latex, or water-based, paint. Latex is a rubber that comes from a tree. When they were first manufactured, latex paints contained this natural rubber. Modern paints, however, use synthetic rubber. Synthetic latex includes acrylics, polyvinyl acetate, and styrene butadiene. Another less popular paint is oil-based. Oil-based paints use natural or synthetic resins derived from oil as their binders. Two common natural oil binders are linseed and tung. Synthetic oils are made from petroleum and include alkyds and polyurethanes.

> **NOTE**
> Many people are allergic to latex. Since modern latex paints do not contain latex, they will not cause allergic reactions in these individuals.

- *Solvents* – Solvents dissolve the binder and make it liquid. They also hold the pigment in suspension, keeping it evenly distributed throughout the paint. When paint is applied to a surface, the solvent dries, leaving the film of binder and pigment. Binders and pigments are referred to as paint solids. Synthetic latex paint uses water as its primary solvent. It also uses other solvents, but in very small amounts. Oil-based or alkyd-based paints use other solvents, such as mineral spirits. Water-based paints allow water to pass through, even after drying, but oil-based paints do not. As a result, oil-based paints can crack and peel if applied over moist surfaces.
- *Additives* – The paint industry spends a great deal of money researching how to improve its products. Additives are used to improve the characteristics of the paint. Additives include wetting agents, which help the paint to spread easily, and drying agents, which speed the paint's drying time.

GOING GREEN

Volatile Organic Compounds

Volatile organic compounds (VOCs) are gases that are released from various substances. Some of these substances are found in certain paints, solvents (such as mineral spirits), varnishes, and waxes. The United States Environmental Protection Agency (EPA) and many state agencies regulate the use of these products. There are many low-VOC products available, so you many never be exposed to substantial amounts of VOCs. To determine if the product you are using emits VOCs, check the manufacturer's literature and the product material safety data sheet.

VOCs can be hazardous to your health. If you are using a product that emits VOCs, you need to make sure there is adequate ventilation in the work area. You may need to use a respirator. Short-term exposure to VOCs can cause eye, nose, and throat irritation, headaches, loss of coordination, and nausea. If you experience any of these symptoms, leave the work area and seek medical assistance, if necessary. Ventilate the work area before returning to work. A VOC monitor may be needed to measure the amount of VOCs in the work area. Long-term exposure can damage the liver, kidneys, and central nervous system.

The paint that you use the most will probably be acrylic latex. This paint has a number of advantages. It has minimal odor, good coverage, is easy to work with, is fast drying, cleans up with plain water, and is available in an unending variety of colors. Acrylic latex paints are very safe to use, but you should always check the product literature and MSDS for the personal protective equipment you must wear. One common adverse reaction some people have to latex paint is contact dermatitis (inflammation of the skin).

Textured paints are an easy way to add interest to a room. Sometimes surfaces are textured with a compound similar to joint compound before they are painted. This method will be discussed later in this module. Another simpler method is to add texturing materials to paint before it is applied to a surface. Some common texturing materials include sand, clay, vermiculite, and polystyrene. Depending on the desired effect, the texture of these materials may be fine, coarse, or anything in between. Texturing a surface is an easy way to conceal imperfections.

Epoxy paints are often used in hospitals, schools, and restaurants where surfaces must be durable, resistant to chemicals, and washable. Epoxy paints are typically used on block or concrete, but they are also formulated for drywall. Some epoxy paints are single-component paints, while others have two components. Single-component epoxy paints have the convenience of being premixed and, when stored properly, they have a long shelf life. Two-component epoxy paints must be thoroughly mixed before use. They have a **potlife** of only a few hours.

3.0.0 ♦ TOOLS

The tools needed to apply interior finishes on drywall are determined by the desired result. A variety of power and manual tools are typically used to complete a job. These tools include paint mixers, paint and texture sprayers, trowels, brushes, and rollers.

INSIDE TRACK

Paint Sensitivity

Even though latex paint does not contain latex, it can still cause allergic reactions in sensitive individuals. This is true of any material. The signs and symptoms of any allergic reaction range from mild (a rash and runny nose) to life threatening (difficulty breathing). Never ignore these symptoms just because you think latex paint is safe.

3.1.0 Paint and Compound Mixers

Paint and compound mixers are used to mix paint, and to mix water and texturing materials into powder compounds. Some mixers are specially designed to mix compounds, paints, or both. Some of these units are self-contained models, such as the mixer shown in *Figure 1*. It can be used to mix mortar, grout, plaster, and paint. Other units are attachments used with power drills.

When paint sits for long periods, the solids in it settle to the bottom of the container, including any texturing material. Depending on the paint's formula, this process can take several weeks or longer. However, when paint contains heavy texturing material, it can settle in a matter of hours. Check the manufacturer's instructions to determine a paint's mixing requirements. Paints that have been recently delivered to a site may not require any mixing if they were mixed by the vendor.

When mixing paint, place the paddle into the container of paint before turning on the mixer. Start at the bottom of the container, and run the mixer in a slow up and down motion to uniformly mix the solids with the liquids. To prevent paint splatters, always keep the paddles beneath the surface of the paint. Also, try to avoid overly vigorous mixing, which can lead to air entrapment in the paint. When this happens, let the paint sit for a few minutes. This will allow the air bubbles to rise to the top.

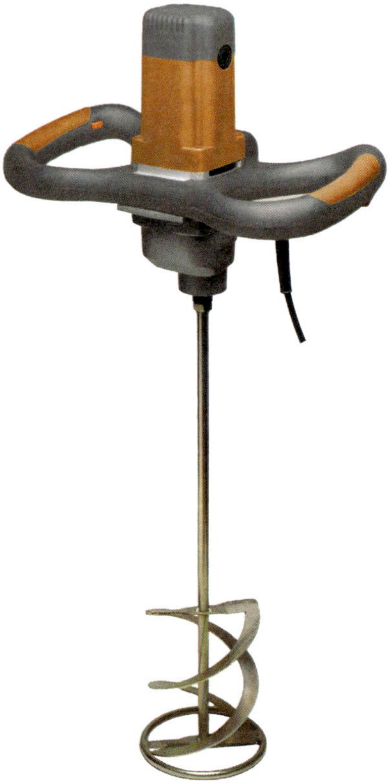

206F01.EPS

Figure 1 ◆ Thirty-gallon mixer.

Mixers are used on compounds in which powders are mixed with water. They are also used whenever texturing material is added to a compound. Texturing material is usually perlite, vermiculite, or polystyrene. These materials are very light in comparison to the **viscosity** of the compound, so they won't settle out.

Many mixers are designed to minimize air entrapment in compounds. This is important because the compounds are thick, and air, which is light, has difficulty rising to the top. When a compound containing entrapped air is applied, the air bubbles may reach the surface before the compound dries. When the bubbles pop, craters or pits are formed on the surface of the compound. A few of these flaws are no problem, but a finish covered with them is unattractive and will need to be sanded and refinished.

When using a mixer, follow these guidelines:

Step 1 Make sure that the mixer is clean before you use it. Dirt, dust, grease, or dried paint or compound can contaminate the paint or compound you are mixing.

Step 2 Put the mixer blade into the container before turning it on.

Step 3 When mixing powder compounds, sift the compound into clean water and mix it by hand. Let the mixture sit for 10 to 15 minutes to thicken. Then add more water, and mix it in with the mixer until the desired consistency is reached.

WARNING!

The dust from powder compounds can cause respiratory and eye irritations. Always use safety goggles and a respirator when handling or working with these powder compounds.

Step 4 When mixing, move the blade slowly across the bottom of the container to thoroughly mix the contents.

Step 5 To prevent air entrapment when mixing compounds or paints, avoid moving the blade in a vigorous up and down motion.

3.2.0 Paint Sprayers

With paint sprayers, large areas can be painted quickly and easily. Sprayers even make painting corners and ceilings relatively simple. Workers skilled in spraying techniques have a surprisingly small amount of overspray. There are two common types of sprayers. One type uses compressed air to turn the paint into a spray. The other style, called an airless system, uses a pump to pressurize the paint. Both systems use a gun to apply the paint to a surface.

Paint sprayers vary by manufacturer. Always be sure to study the operator's manual before using an unfamiliar piece of equipment. The following are general guidelines for using sprayers:

- Clean all sprayer equipment after use according to the manufacturer's specifications. This usually involves running the correct paint solvent through the sprayer until it is clean. Water is the appropriate solvent for latex paint. Always be sure to clean sprayer equipment properly before storage.
- Before starting any paint job, make sure that the equipment is clean. Check and clean all filters. Wipe dust and dirt from the equipment to avoid contaminating the paint.
- Disconnect the power source before performing any servicing of the equipment. To release any built-up pressure, point the gun into the reservoir and pull the trigger.
- Never point a spray gun at anyone, and never spray another worker with paint.
- Check the manufacturer's instructions to determine the proper viscosity of the paint.
- Practice spray painting on scraps of drywall. Before you start painting, adjust the sprayer pressure and gun aperture until the paint flow is correct.
- Hold the gun so that it is 8 to 12 inches from the surface being painted.
- Position the gun perpendicular to the surface being painted. Move the gun at a rate of speed that will allow for adequate paint coverage.
- Overlap strokes to blend the paint.
- Some textured paints cannot be applied with a sprayer. The texturing material will clog the sprayer. When applying textured paints, always check the paint manufacturer's recommendations before beginning the job.

3.3.0 Spray Texturing Machines

Spray texturing machines are used to apply a texture to a surface before the surface is painted. The spray texturing gun with hopper shown in *Figure 2* is designed to work with an air compressor. The texturing material is placed in the hopper, which has a small hole at the bottom where the hopper joins the gun. This opening allows material to be

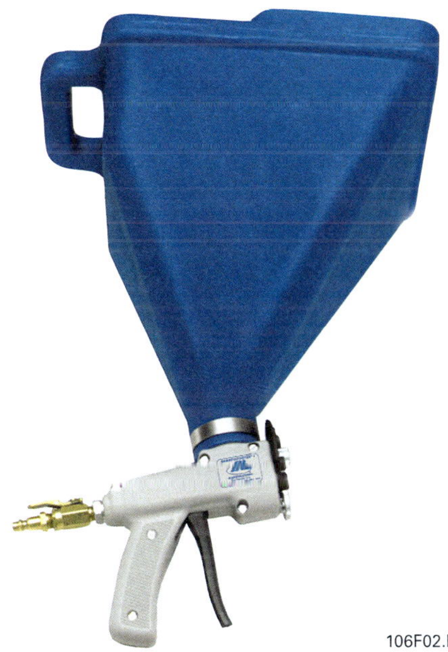

106F02.EPS

Figure 2 ◆ Hopper gun.

GOING GREEN

Reusing Solvent

Used solvent may be cleaned and reused. Place used solvent into a container, cover it, and allow it to sit undisturbed for several hours. The paint solids will sink to the bottom of the container. The top portion of the solvent can be poured off and reused.

fed into the gun. When the trigger is pulled, compressed air rushes past the hopper opening, and texturing material and air are shot out of the gun. The air rushing past the hopper opening also creates a suction that pulls the material into the gun. The gun is controlled with the operator's dominant hand, while the other hand usually supports the hopper.

Another type of portable texture sprayer is shown in *Figure 3*. The texturing material is placed in a canister that is carried on the operator's back. This canister, which holds 2½ gallons of texture, is fed compressed air from an external air compressor. The texturing material is applied with a gun. This design is easier to manage than the hopper gun arrangement, because the operator is free from the need to support the hopper while spraying.

Large jobs require large equipment. The spray texturing machine shown in *Figure 4* has a larger hopper than the one mounted on the gun in *Figure 3*. It also has its own air compressor. For even bigger jobs, there are bigger machines. The one shown in *Figure 5* holds 300 gallons.

Texturing guns need to have an adjustable **orifice** for applying small, fine textures as well as large, coarse textures. Using an orifice that is too small will clog the gun. Using one that is too large will apply too much texturing material.

As with paint sprayers, texturing guns vary by manufacturer. Always be sure to study the operator's manual before using any equipment for the first time. The following are general guidelines for using a texturing machine:

- After using any texturing machine, be sure to clean it according to the manufacturer's specifications. Never store equipment until it has been properly cleaned.
- Before beginning any job, make sure that the equipment is clean. Wipe any foreign matter from the hopper to avoid contaminating surfaces. Substances, such as oil and grease, can prevent paint from bonding to the texturing compound.
- Prior to servicing any equipment, always disconnect the power source.
- Before beginning a job, practice applying the texturing material on scrap drywall.

206F04.EPS

Figure 4 ◆ Self-contained texturing machine.

206F03.EPS

Figure 3 ◆ Portable texturing system.

206F05.EPS

Figure 5 ◆ Three-hundred-gallon texturing rig.

- Hold the gun so that it is 12 to 16 inches from the surface being painted.
- The technique used when applying texturing materials depends on the desired result. Once a method has been selected, it must be maintained it until the job is finished.
- When texturing a ceiling, use a gun that has an angled orifice, if possible.

3.4.0 Hand Tools

Brushes and rollers are impractical for painting commercial areas, although they may find limited use for doing **cut-in** work around fixtures and touching up small areas. Because of their size, commercial spaces are usually spray painted. This allows the job to be completed quickly and the paint to be applied evenly. Even when applying a faux finish to a surface, the base coat is usually applied with a sprayer. Hand tools, however, are used to apply the detail coats. Hand tools are also used to apply patterns to texturing compounds (*Figures 6* through *8*).

A number of hand tools are available for texturing walls and applying faux finishes. These hand tools are listed in *Table 1*.

4.0.0 ◆ APPLICATION

A finish coat on walls and ceilings is one of the first things that people notice when they walk into a room. Whether the finish is smooth or textured, it should look professional, neat, and attractive. Applying the final finish to a surface is only one step in the entire process. Finish work begins with surface preparation.

4.1.0 Surface Preparation

Surface preparation can be the most time consuming part of finish work. To properly prepare a surface for finish work, you will need to know how the surface is supposed to appear in the final product. Surfaces finished with texturing compound do not need to have smoothly sanded compound on every joint and fastener head. On the other hand, surfaces finished with smooth paint show every bump, depression, and ridge. This is especially true when the final finish is done with a dark-colored, high-gloss paint, which magnifies any defects.

Paying close attention to details is important in any finishing job, even when textured materials or paints are applied to a surface. Before you apply a finish to any surface, always carefully examine the surface for imperfections that will

206F06.EPS

Figure 6 ◆ Stipple brush.

206F07.EPS

Figure 7 ◆ Tandem texture brush.

206F08.EPS

Figure 8 ◆ Foam texture roller.

Table 1 Manual Tools

Tool	Description	Purpose	Uses
Stucco brush	Wide, stiff-bristled brush with a handle.	Creates a variety of textures from stippling to swirls.	Texturing compound
Texture brushes	Stiff-bristled brushes with or without a handles. May be tandem-mounted to cover a larger area.	Creates a variety of textured patterns.	Texturing compound
Texture rollers	Rollers come in a variety of materials and textures depending on the finish desired. Rollers slip onto a holder that has handles. The handle can be attached to an extension for reaching overhead.	Creates a variety of textures.	Texturing compound
Wipe-down blade	A wide, hardened steel blade with rounded corners to prevent gouging. May have a long or short handle.	Wipes excess material from a surface. Helps clean walls and floors after application of texturing materials.	Texturing compound
Paint roller and pan	Rollers come in a variety of materials and textures to suit the type of paint used and the finish desired. They also come in a variety of widths. Rollers slip onto a holder that has a handle. The handle can be attached to an extension for reaching overhead. Pans come in various sizes.	Applies paint.	Smooth and textured paint
Flat-blade knife	Short-handled metal or plastic blade that comes in varying widths.	Applies texturing material; also used for troweled finishes.	Texturing compound
Sponges	Synthetic or natural (sea) sponge.	Applies free-form texturing finishes.	Texturing compound and faux finishes
Rags and paper	Clean cloths and soft or stiff paper.	Applies free-form texturing finishes.	Faux finishes
Whisk broom	Stiff bristled brush.	Creates a variety of textures from stippling to swirls. Similar to a stucco brush, but can be used to produce a bolder brush pattern.	Texturing compound
Trowel	Rectangular- or triangular-shaped metal or plastic blade with a handle. Edges may be smooth or have a pattern.	Applies texturing materials; also used to achieve troweled or knockdown finishes.	Texturing compound

206T01.EPS

show through the final application. Follow these guidelines when you inspect a surface:

- Use a systematic approach. Start at one corner, and work your way around the entire area.
- Understand the degree of surface preparation that is required. Smooth finishes need to be applied to smooth surfaces. Sand any bumps, ridges, and high spots as needed.
- Examine all joints and fastener holes, and make sure that they are covered with enough joint compound and are adequately sanded. Flag any areas that need to be reworked. One way to flag an area is by using colored Post-it® notes. Another way is to use a marker or colored chalk to draw an arrow on the floor (provided the floor is bare).

- Check all corners to be certain that the bead is adequately covered.
- Look for any pits, gaps, depressions, shrinkage, or flaking in the joint compound, and repair as needed.
- Inspect the drywall surface for cracks, and repair as needed.
- When a skin coat is used, examine the entire surface, and make sure that all areas have been well coated.
- When necessary, clean all dust from surfaces before applying any texturing material, primer, or paint.
- Cover any fixtures that will not be coated, such as windows and electrical fixtures.

4.2.0 Primer

Primer is a paint that is applied to a bare surface to help prepare it for the final coat. Primer fills in holes, blocks some stains to prevent them from bleeding through the final coat, and provides a better surface than drywall paper for adhering the top coat. Bare drywall is usually primed with a water-based primer. However, when texturing will be applied, the primer will be oil-based. Oil-based primer protects the drywall from the water in the texturing compound. Surfaces primed with an oil-based primer can be painted with either a latex or an oil-based top coat. Surfaces primed with a latex primer may only be painted with a latex top coat.

Not all surfaces require primer, so check the project specifications to find out if one is required. If this is not specified, follow the manufacturer's instruction for the product being used as the finish coat. Contact the manufacturer if necessary.

Latex primer is usually applied at temperatures between 50° and 90°. Oil-based primers can be applied at temperatures as low as 40°. Primers that are applied with a roller should not be thinned. Primers that are sprayed on are often thinned. Check the manufacturer's literature for instructions. Never thin primer more than directed by the manufacturer.

4.3.0 Spray Painting

A skilled operator can spray paint an area quickly and efficiently, but an unskilled operator can make a mess. Before you use any spraying equipment, first study the operator's manual, and then practice on scrap pieces of drywall. Make sure that the machine is clean and in good working order before you load it with paint.

Before you begin painting, calculate the amount of paint needed to complete the job. To do this, you must determine the surface area of all surfaces to be painted. Follow these steps to determine the surface area:

Step 1 Measure the length and height of each wall, and then multiply these two numbers to determine the surface area. For the room shown in *Figure 9*, the surface areas are determined as follows:

Wall A
Length × height = 34' × 12' = 408 sq ft

Wall B
Length × height = 10' × 12' = 120 sq ft

Wall C
Length × height = 7' × 12' = 84 sq ft

Wall D
Length × height = 7' × 12' = 84 sq ft

Wall E
Length × height = 10' × 12' = 120 sq ft

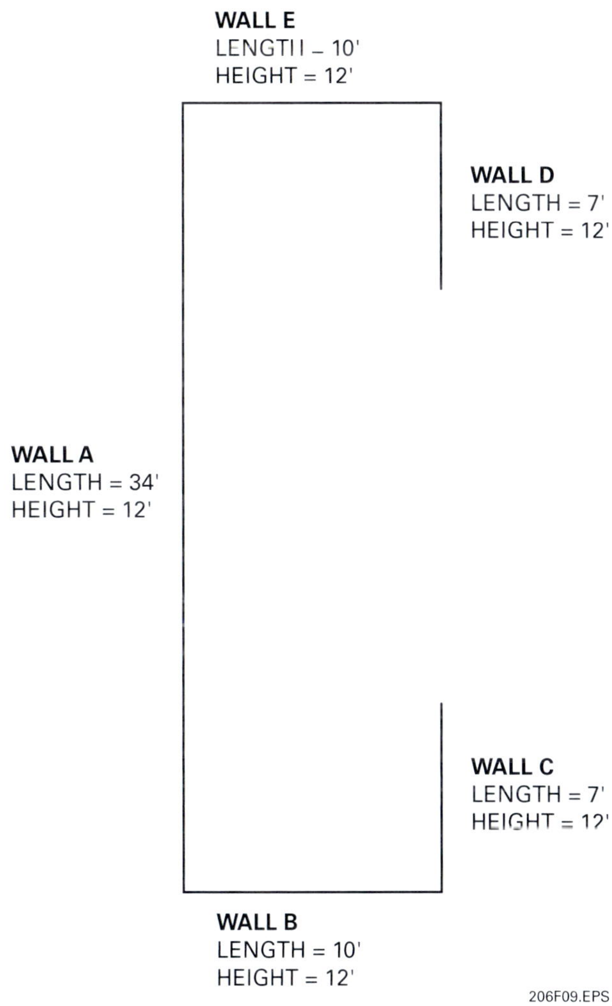

Figure 9 ◆ Calculating paint needs.

Step 2 Add together the areas of all surfaces. For example, using the areas calculated in Step 1, the sum is as follows:

408 + 120 + 84 + 84 + 120 = 816 sq ft

Step 3 Calculate the area for each opening in the walls. For example, assume that Wall A has a 10' by 4' door and two 6' by 4' windows. Their areas are calculated as follows:

Door = 10' × 4' = 40 sq ft

Window One = 6' × 4' = 24 sq ft

Window Two = 6' × 4' = 24 sq ft

Step 4 Add the areas of all openings together. Using the areas calculated in Step 3, the sum is as follows:

40 + 24 + 24 = 88 sq ft

Step 5 Subtract the sum of the openings from the surface area as follows:

816 – 88 = 728 sq ft

Step 6 Divide the surface area by the manufacturer's stated coverage per gallon. One gallon of latex paint typically covers 400 to 450 square feet. Using 400 square feet per gallon, the calculation is as follows:

728 sq ft ÷ 400 sq ft per gallon = 1.8 gallons

Step 7 The minimum amount of paint you need to paint the room shown in *Figure 9* is 1.82 gallons. Because it is impossible to spray without wasting some paint, an allowance for waste should be added to this calculation. Allowing ten percent for waste, the calculation is as follows:

1.82 gallons × 1.10 percent = 2.002 gallons (usually rounded to 2 gallons)

The above calculations are for painting smooth finishes. For painting textured finishes, up to 50 percent more paint will be needed.

The paint may need to be thinned before it can be used in the sprayer. For information about thinning the paint, check the operator's manual for the sprayer machine and the paint manufacturer's literature. Never thin the paint more than the manufacturer's recommendation.

Follow these guidelines when using a spray painting machine:

- Adjust the machine pressure and gun orifice so that enough paint is released when the trigger is pulled.

INSIDE TRACK

Protect Yourself

Spray painting releases paint droplets into the air that can easily be inhaled. Check the paint MSDS to determine whether you need to wear any respiratory protection. A MSDS will state the minimum protective measures that need to be taken when using a product. These measures are for your protection, so be sure to follow them.

- Hold the nozzle perpendicular to and 8 to 12 inches from the surface being painted. Start moving the gun across the test surface, and then pull the trigger to release the paint.
- Avoid moving your arm in a pronounced arc. This will release paint droplets into the air and will result in too little paint on the ends of the sweep stroke.
- Slowly sweep the gun back and forth over an area. Continue overlapping sweeps until the coating is uniformly applied.
- Adjust the speed of application and/or the paint flow until adequate coverage is achieved without any drips or runs.
- Release the trigger at the end of a sweep to avoid over-painting the ends.
- Adjust your equipment and application technique so that you can paint an area with a minimum number of sweeps.
- Clean up oversprays and spills as soon as possible. Wet paint is easier to clean up than dry paint.

5.0.0 ◆ SPECIALTY FINISHES

Specialty finishes are an easy way to make a room more attractive. The most common finish used today is texturing. Texturing may be applied in one of two ways—by using a texturing compound on the surface before it is painted, or by using a textured paint. Texturing a surface takes a little more time than using a textured paint, but it will produce a more distinctive look. For an even more unique effect, a faux finish can be applied to a surface. Faux finishes use paint to make a surface look like marble, stone, leather, or other luxury material.

5.1.0 Texturing Compound

Texturing compounds are applied to a surface before it is painted. These compounds may be sprayed on, or they may be applied by hand

and worked into the desired texture. Another option is to apply the compound with a sprayer and then work it into the desired texture by hand. Texturing compounds are similar to joint compounds, and some joint compounds may be used as texturing compounds. Texturing materials can come in powder form or as a ready-mixed product. Some compounds are smooth, while others contain some type of aggregate to give them more texture. The aggregate is usually perlite, vermiculite, polystyrene, or Styrofoam™. It may be fine, medium, or coarse in texture. Before starting any texturing job, it's a good idea to practice first on pieces of scrap drywall.

5.1.1 Spray-On Textures

Texturing compounds can be splattered onto a surface by machine and then left to dry so that the texture is raised. This type of texture is called popcorn (*Figure 10*). Because popcorn texture is easily damaged, it is usually used on ceilings rather than on walls. Popcorn texturing can transform a smooth surface into a very rough one, and is commonly referred to as simulated acoustical spray.

Another popular texture is orange peel (*Figure 11*). This texture is created by using a thinner compound and lower air pressure than are used for applying the popcorn texture. This technique allows the globules of compound to spread a bit as they dry, creating an attractive appearance, somewhat like an orange peel.

A popular texture called knock-down (*Figure 12*) is made using the same compound as orange peel, with or without the aggregate. First, the compound is sprayed on and allowed to set for 15 minutes or more. Then, a steel blade is manually dragged across the surface, knocking down the peaks. This texture takes two or more workers to complete—one to spray the compound onto the surface, and the others to do the knock-down work.

Follow these guidelines when applying texturing compounds with a sprayer:

- Spraying can be very messy. Be sure to dress for the job.
- Mask off or cover anything that should not be sprayed.
- When applying knock-down texture, timing is important. Adjust your speed so that the sprayed area can be worked before it becomes too dry.
- Clean up oversprays promptly. To remove dried compound, spray it with water, allow it to sit for a minute or two, and then scrape it with a putty knife.

206F10.EPS

Figure 10 ◆ Popcorn texture.

206F11.EPS

Figure 11 ◆ Orange-peel texture.

206F12.EPS

Figure 12 ◆ Knock-down texture.

5.1.2 Other Textures

Texturing compound can be worked in numerous ways to achieve an attractive finish. Tools used can include brushes, sponges, trowels, rags, or blades. When the surface texture will be worked manually, the compound may be applied manually or with a sprayer. For a light texture, the compound should be about ⅛-inch thick. Deeper textures will require a thicker layer. Many textures are manually applied and worked into the final surface. Others are sprayed onto the surface and then manually worked to achieve the desired texture.

The compound is applied and then distributed fairly evenly across the surface with a wide blade. It is not necessary to smooth the compound—it should be distributed in an attractively rough pattern. This finish could be the final finish, or the compound may be worked with a roller, brush, sponge, rag, or other tools until the desired finish is achieved.

Figure 13 shows various types of textured finishes. The light stippling is created by gently pressing a stippling brush into the wet compound. The deeper stipple uses the same brush,

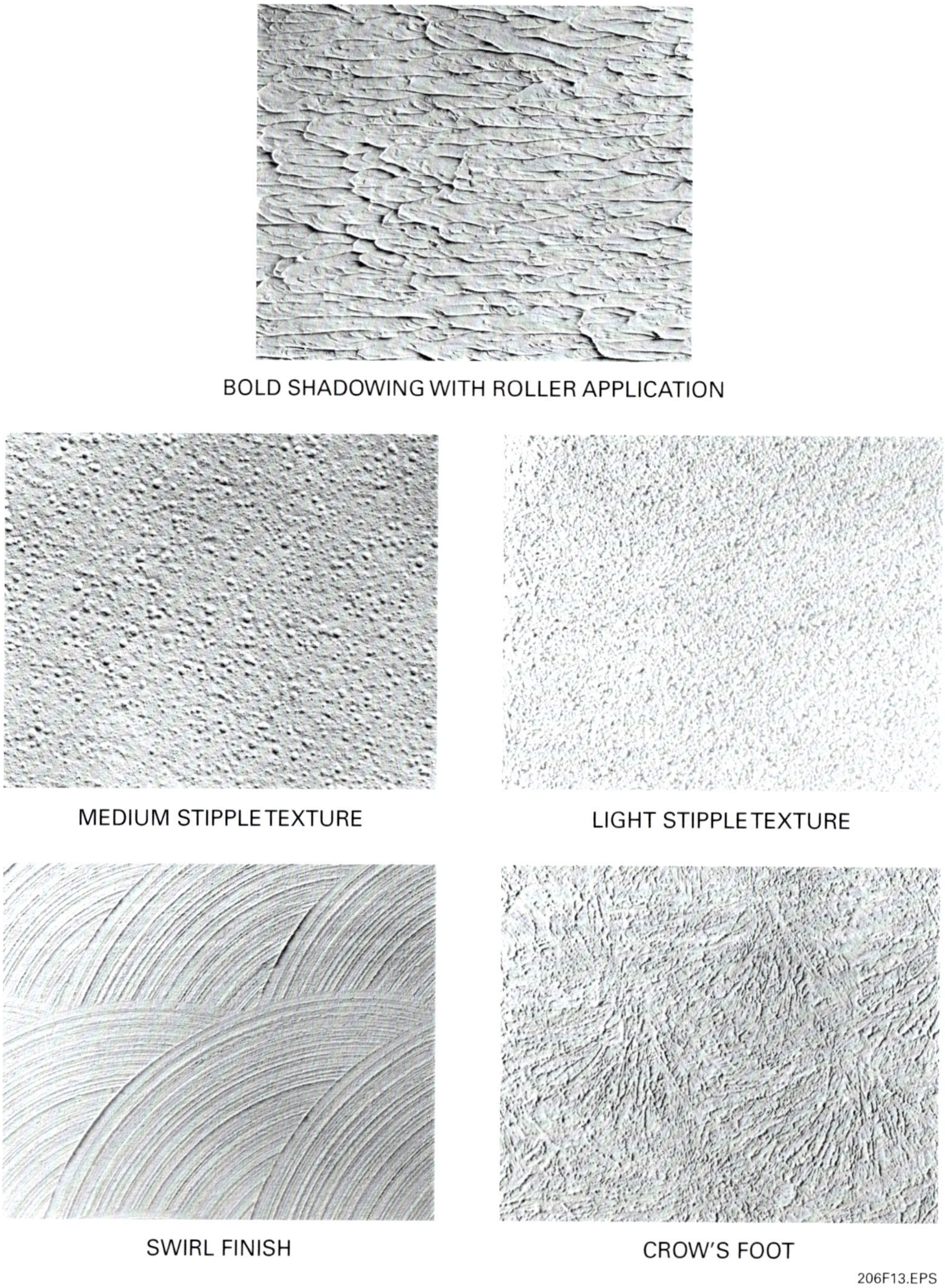

Figure 13 ◆ Examples of textured finishes.

but it is pressed more firmly into the compound. To create the swirl finish, a stiff-bristled brush is worked across the wet compound in a fan-like motion. Bold shadowing is achieved by running a roller across the wet compound, while the crow's foot pattern is made using a textured roller.

5.2.0 Textured Paints

Textured paints are applied in the same manner as smooth paints. Most may be rolled, brushed, or sprayed on. Some types of textured paints require periodic mixing during application to redistribute the texturing material. Check the manufacturer's instructions for details.

5.3.0 Faux Finishes

Faux finishes are made by applying layers of paint on a surface. Faux finishes take more effort than other types of finishes, so they are often used in a limited area. Two popular faux finishes are marbling and Venetian plaster. Marbling uses a number of layers of paint that are applied in a special way to give the final surface the appearance of marble. This finish is often used on columns. Venetian plaster simulates a polished plaster finish popular in Europe. Before using these faux finish techniques on the job, it's a good idea to perfect your skills on scraps of drywall.

5.3.1 Marble Faux Finish

Applying a marble faux finish to a surface takes time and patience. For this finish, you will need the following supplies:

- Latex satin paint (one base color and at least two others shades)
- Latex satin paint for veining (white, gray, green, or another shade of the base color)
- Clear polyurethane varnish or wax
- Paint roller and holder and paint pan
- Natural sea sponge
- Pointed artist's brush

Natural marble contains many shades of color. The more shades of color you use, the more realistic the finish will look, but use the paint sparingly. The following is a general procedure for applying a marbled faux finish. As you gain experience, you will likely develop your own technique.

Step 1 Apply the base coat over the entire surface with a damp sponge, roller, or sprayer (*Figure 14*). This coat does not need to cover the surface completely so that none of the surface color shows through. When rolling the paint, be sure to smooth out all roller marks. Allow this coat to dry.

Step 2 Alternate applying the two lighter colors with a damp sponge until the desired effect is obtained (*Figure 15*). Using a light touch, apply the sponge directly to the surface, and then remove it. Be careful not to drag the sponge across the surface. Some drywall technicians allow this coat to dry before moving on to the next step. Others proceed immediately to the next step.

Step 3 Using an artist's brush and the veining paint, lightly paint veins on the surface. Be sure to vary the appearance and size of the veins (*Figure 16*). The key to this technique is to use a light hand. Veins should be delicate. Allow this coat to dry.

206F14.EPS

Figure 14 ◆ Marble base coat.

206F15.EPS

Figure 15 ◆ Marble second coat.

206F16.EPS

Figure 16 ◆ Marble detail coat.

Step 4 To seal the coating, apply a faux finish varnish or wax over the surface according to the manufacturer's instructions.

5.3.2 Venetian Plaster

Creating a Venetian plaster faux finish (*Figure 17*) starts with Venetian plaster or Venetian plaster paint. To achieve this faux finish, two or three layers of plaster or paint are applied to a surface with a trowel. After the paint or plaster has dried, the surface is polished. This polishing, also called burnishing, is what gives the plaster its glossy finish.

Venetian plaster paint is thicker than regular latex paint. It is specially formulated to achieve the desired surface. Be sure to stir the paint thoroughly before use. Also, wipe the application blade from time to time to avoid contaminating the surface with dried plaster. Practice on scrap material until you achieve the desired result. To create this faux finish, you will need the following supplies:

- Venetian plaster or Venetian plaster paint
- Venetian plaster or paint top coat
- Steel trowel or blade with rounded corners
- Clean cloth, power buffer, or power sander with 600-grit sandpaper

The following is a general procedure for applying a Venetian plaster faux finish. Be sure to follow the manufacturer's application instructions for the product being used.

Step 1 Apply plaster or paint to the surface with a trowel or blade. Hold the tool so that the blade is at a 15- to 45-degree angle to the surface. Spread the plaster or paint thinly with a rough finish. Apply the coating in a random X-pattern using long and short strokes. Smooth out any obvious high spots, but leave the surface rough. This coat does not need to cover the surface completely, since another coat will be applied. Allow this coat to dry.

Step 2 Apply the second layer of plaster in a similar manner to the first, but hold the tool so the blade is at a 45- to 90-degree angle to the surface. Make sure that this coat completely covers the surface. Allow this coat to dry.

Step 3 If desired, apply a third layer of plaster, as you did with the second coat, and allow this coat to dry.

206F17.EPS

Figure 17 ◆ Venetian plaster faux finish.

Step 4 Using a moistened rag, rub the surface of the plaster until a glossy sheen appears. Instead of a rag, you can use a power buffer or sander to lightly polish the plaster.

> **NOTE**
> Some products recommend applying the top coat to the plaster before it is polished.

Step 5 Apply the protective topcoat to the plaster.

6.0.0 ◆ SAFETY

As a drywall technician, it is your responsibility to follow good safety practices at work, and to report any unsafe conditions to your employer. Workers owe it to themselves, their families, and their co-workers to report any unsafe conditions. The best way to prevent injuries is to use common sense. Good general housekeeping helps to protect workers from accidents and injuries. Always be aware of your surroundings. When you see a potential hazard, fix it. If you can't fix it, report it to your supervisor.

Consult the MSDS for each product you are using. Don't assume that every product is harmless just because you usually use safe, stable products. Be aware of the stability of all materials that you handle. Take the steps needed to guard against fires and explosions. Collect rags soaked with solvent and place them in an approved container. Never smoke or eat near a painting operation.

Personal protective equipment is designed to protect workers from injury. Many on-the-job injuries occur because workers fail to use the proper personal protective equipment. When working with chemicals, be sure to check the MSDS for specific information about the proper types of PPE to wear. Always check the MSDS before using a product.

The personal protective equipment usually used when applying paint includes the following:

- Eye protection
- Gloves
- Respiratory protection
- Protective clothing

WARNING!

Don't take chances with your safety. Use your personal protective equipment every time the job requires it.

6.1.0 Hazardous and Nonhazardous Waste Disposal

Most of the paints you use will be classified as nonhazardous waste. This means that the paint can be disposed of in accordance with the policy of the employer. These policies are usually based on local laws about waste disposal. Be sure to follow your company's policies about nonhazardous waste disposal.

Some paints and solvents are classified as hazardous waste. These products are dangerous or potentially dangerous to human health or the environment. Hazardous waste must be disposed of in accordance with the Environmental Protection Agency (EPA) *Resource Conservation and Recovery Act (RCRA) 40 CFR 261*, or with local law. Local law can be stricter than the RCRA, but it cannot be more lenient. The MSDS will tell you how to dispose of a product.

If a hazardous material gets spilled in an area, check the MSDS for cleanup instructions. The EPA may need to be informed of the spill, depending on the material and the size of the area affected. Some states may have more stringent rules about handling hazardous materials and cleaning up spills. Check with your supervisor if you have any questions.

Review Questions

1. One of the greatest advantages of texturing a surface is that it _____.
   a. hides imperfections
   b. looks like marble
   c. is applied with rags
   d. looks natural

2. The paint component that determines the finish color is the _____.
   a. additive
   b. pigment
   c. binder
   d. solvent

3. Latex paint _____.
   a. takes a long time to dry
   b. is available in a limited number of colors
   c. has a low odor
   d. has poor coverage

4. The potlife for epoxy paints after they are mixed is _____.
   a. a few hours
   b. several days
   c. a month or more
   d. usually more than three days

5. To mix powder compounds, _____.
   a. always use a mixer
   b. initially sift the powder into the water and mix by hand
   c. add enough powder to achieve the desired consistency
   d. sift the powder into a clean container and then add clean water

6. To clean a paint sprayer machine after using latex paint in it, you should circulate _____ through it.
   a. water
   b. bleach
   c. detergent
   d. turpentine

7. The most time consuming part of finish work is _____.
   a. careful paint application
   b. good surface preparation
   c. choosing high quality application tools
   d. applying texturing compound neatly

8. Before applying primer, it is very important to _____ the surface.
   a. clean dust from
   b. slightly moisten
   c. vigorously scrape
   d. lightly sand

9. You should always prime a drywall surface before applying a finish coat.
   a. True
   b. False

10. Latex primers should be applied at temperatures between _____ and 90°.
    a. 30°
    b. 50°
    c. 70°
    d. 72°

11. When painting with a sprayer, adjust your equipment so that you can paint the area with _____.
    a. an arc-like motion
    b. a minimum number of sweeps
    c. as much paint as possible
    d. no protective gear

12. The aggregate in a texturing compound may be _____.
    a. latex
    b. acrylic
    c. powder
    d. coarse, medium, or fine

Review Questions

13. To achieve an orange peel textured finish, you use the same compound as popcorn textured finish but you _____.
    a. thin it with water and use lower air pressure
    b. manually apply it
    c. spray it with water after it is applied
    d. thin it and then apply it with a brush

14. To remove dried texturing compound from a floor, you should _____.
    a. pound it with a mallet
    b. spray it with water, then scrape it with a putty knife
    c. scrub it with mineral spirits
    d. use a power washer

15. To achieve a Venetian plaster finish, you can use _____.
    a. two layers of different colored paints
    b. a small brush to paint details
    c. a specially formulated paint or plaster
    d. Venetian rubbing compound

Summary

Finish work is what gives modern commercial buildings their distinctive look. Many of the products used to finish surfaces can be applied quickly and easily. At the same time, they can dramatically alter the appearance of a building. When working on a job, keep in mind that a good finish job is a dynamic process that starts with professionally placed drywall and continues on until the final coat is applied to a surface.

Notes

Trade Terms Introduced in This Module

Cut-in: A technique used to manually paint around a fixture with a brush or roller.

Potlife: The period of time in which a mixed, two-component epoxy paint must be used.

Viscosity: The thickness of a fluid.

Additional Resources and References

Additional Resources

This module is intended to be a thorough resource for task training. The following reference works are suggested for further study. These are optional materials for continued education rather than for task training.

Drywall: Professional Techniques for Great Results, 2002. Myron R. Ferguson. Newtown, CT: Taunton Press.

Installing & Finishing Drywall, 1998. William Spence. New York, NY: Sterling Publishing Company.

Figure Credits

Photo courtesy of Northern Tool + Equipment, Burnsville, MN, 206F01

Marshalltown Company, 206F02, 206F03, 206F06, 206F07

The Stanley Works, 206F04

Copyright 2008, All-Wall Inc. All rights reserved–used with permission, 206F05

Kraft Tool Company, 206F08

Pictures supplied and © by Universal Spray & Drywall Inc., 206F10–206F12

USG Corporation, 206F13

Zinsser, 206F14–206F16

Modern Masters, 206F17

NCCER CURRICULA — USER UPDATE

NCCER makes every effort to keep its textbooks up-to-date and free of technical errors. We appreciate your help in this process. If you find an error, a typographical mistake, or an inaccuracy in NCCER's curricula, please fill out this form (or a photocopy), or complete the online form at **www.nccer.org/olf**. Be sure to include the exact module ID number, page number, a detailed description, and your recommended correction. Your input will be brought to the attention of the Authoring Team. Thank you for your assistance.

Instructors – If you have an idea for improving this textbook, or have found that additional materials were necessary to teach this module effectively, please let us know so that we may present your suggestions to the Authoring Team.

NCCER Product Development and Revision

13614 Progress Blvd., Alachua, FL 32615

Email: curriculum@nccer.org
Online: www.nccer.org/olf

❑ Trainee Guide ❑ Lesson Plans ❑ Exam ❑ PowerPoints Other ______________________

Craft / Level: Copyright Date:

Module ID Number / Title:

Section Number(s):

Description:

Recommended Correction:

Your Name:

Address:

Email: Phone:

Glossary of Trade Terms

Acoustical materials: Types of ceiling tile, plaster, and other materials that have high absorption of sound waves.

Acoustics: A science involving the production, transmission, reception, and effects of sound. In a room or other location, it refers to those characteristics that control the reflection of sound waves and the sound reception in the area.

ADA Requirements: The Americans with Disabilities Act of 1990 requires the use of accessible design methods for the construction of public accommodations such as lodging, public parks, schools, and restaurants.

As-built: Revised drawings of an installation that show the changes made during installation.

A-weighted decibel (dBA): A single number measurement based on the decibel but weighted to approximate the response of the human ear to the intensity of sound sources.

Beams: Loadbearing horizontal framing elements supported by walls or columns and girders.

Blocking: C-shaped track, break shape, or flat strap material attached to structural members, flat strap, or sheathing panels to transfer shear forces.

Building codes: Codes published by state and local governments to establish minimum standards for various types of interior and exterior construction.

Callouts: Markings or identifying tags describing parts of a drawing; callouts may refer to detail drawings, schedules, or other drawings.

Ceiling panels: Any lay-in acoustical board that is designed for use with an exposed grid mounting system. Ceiling panels normally do not have finished edges or precise dimensional tolerances because the exposed grid mounting system provides the trimout.

Ceiling tiles: Acoustical ceiling boards that are nailed to, cemented to, or suspended from a concealed grid mounting system. The edges are often kerfed and cut back.

Civil drawings: A drawing that shows the overall shape of the building site. It is also called a site plan.

Clip angle: An L-shaped piece of steel (normally having a 90-degree bend), typically used for connections.

Cold-formed steel: Sheet steel or strip steel that is manufactured by press braking blanks that have been sheared from sheets or cut length of coils or plates, or by continuous roll forming of cold- or hot-rolled coils of sheet steel.

Compressive strength: Refers to a material's ability to support weight.

Contour lines: Imaginary lines on a site plan/plot plan that connect points of the same elevation. Contour lines never cross each other.

Control joint: A trim piece used with stucco and exterior insulation and finish systems to relieve stress in the coating due to thermal expansion and contraction.

C-shape: A cold-formed steel shape used for structural and nonstructural framing members consisting of a web, two flanges, and two lips (edge stiffeners).

Curtain wall: A light, nonbearing exterior wall attached to the concrete or steel structure of a building.

Cut-in: A technique used to manually paint around a fixture with a brush or roller.

Diaphragm: A floor, ceiling, or roof assembly designed to resist in-plane forces such as wind or seismic loads.

Diffuser: An attachment for duct openings in air distribution systems that disperses the air in wide flow patterns. In lighting systems, it is an attachment used to redirect or scatter the light from a light source.

Elastomeric: Having the elastic properties of rubber.

Elevation view: A drawing giving a view from the front, rear, or side of a structure.

Fissured: A ceiling panel or ceiling tile surface design that has the appearance of splits or cracks.

Flashing: Thin strips of material that are placed above openings in a structure to divert water away from the opening.

For-fill: A wall to be filled with some type of material, such as concrete or insulation.

Frequency: Cycles per unit of time, usually expressed in hertz (Hz).

Glossary of Trade Terms

Girders: Large steel or wooden beams supporting a building, usually around the perimeter.

Header: A horizontal structural framing member used over floor, roof, or wall openings to transfer loads around the opening to supporting structural framing members.

Hertz (Hz): A unit of frequency equal to one cycle per second.

Isometric drawing: A three-dimensional type of drawing in which the object is tilted so that all three faces are equally inclined to the picture plane.

Joists: Horizontal members of wood or steel supported by beams and holding up the planks of floors or the lathes of ceilings. Joists are laid edgewise to form the floor support.

Knurled: A series of small ridges on pins used to provide a better gripping surface on metal and plastic.

Landscape drawings: A drawing that shows proposed plantings and other landscape features.

Lateral: Running side to side; horizontal.

Lips: That part of a C-shape framing member that extends from the flange as a stiffening element that extends perpendicular to the flange; also called edge stiffener.

Mil: A unit of measurement equal to 1⁄1,000 of an inch.

Panelization: The process of assembling steel-framed walls, joists, or trusses before they are installed in a structure. Roll-formers normally cut studs to within 1⁄8-inch tolerance. This helps the framer to consistently create straight walls on the panel table that are easy to install in the field.

Permeable: Having pores or openings that permit liquids, such as water, or gases to pass through.

Plan view: A drawing that represents a view looking down on an object.

Plenum: A confined space such as the area between a suspended ceiling and an overhead deck that is used as a return for a heating, cooling, and ventilation systems.

Potlife: The period of time in which a mixed, two-component epoxy paint must be used.

Racking: Being forced out of plumb by wind or seismic forces.

Rim track: A horizontal structural member that is connected to the end of a floor joist.

Riser diagram: A type of isometric drawing that depicts the layout, components, and connections of a piping system.

Schedules: Charts or tables that provide detailed information corresponding to various parts of the drawings.

Sections: Drawings that detail construction techniques or materials.

Shear wall: A wall designed to resist lateral forces such as those caused by earthquakes or wind.

Spandrel: A beam that is an angled support for framing.

Starter strip: A trim piece used with exterior insulation and finish systems to support the insulated panels.

Striated: A ceiling panel or ceiling tile surface design that has the appearance of fine parallel grooves.

Tensile strength: Refers to the flexibility of a material.

Title block: A section of an engineering drawing blocked off for pertinent information, such as the title, drawing number, date, scale, material, draftsperson, and tolerances.

Track: A framing member consisting of only a web and two flanges. Track web depth measurements are taken to the inside of the flanges.

Veneer: A thin sheet of material, such as wood, that is glued to some type of fabricated material.

Viscosity: The thickness of a fluid.

Weep screed: A trim piece used with stucco and other cladding systems to drain water from behind the cladding.

Index

Index

C

O

P

R

T